Computers
as
Assistants

A New Generation of
Support Systems

COMPUTERS, COGNITION, AND WORK

A series edited by:

Gary M. Olson, Judith S. Olson, and Bill Curtis

Computers
as
Assistants

A New Generation of Support Systems

Edited by PETER HOSCHKA

Institute for Applied Information Technology (FIT)
GMD – German National Research Center for Information Technology

CRC Press
Taylor & Francis Group
Boca Raton   London   New York

CRC Press is an imprint of the
Taylor & Francis Group, an **informa** business

First Published by
Lawrence Erlbaum Associates, Inc., Publishers
10 Industrial Avenue
Mahwah, New Jersey 07430

Transferred to Digital Printing 2009 by CRC Press
6000 Broken Sound Parkway, NW Suite 300, Boca Raton, FL 33487
270 Madison Ave, New York, NY 10016
2 Park Square, Milton Park, Abingdon, Oxon OX14 4RN, UK

Cover design by Gail Silverman

Library of Congress Cataloging-in-Publication Data

Computers as assistants : a new generation of support systems
/ edited by Peter Hoschka.
     p. cm.
  Includes bibliographical references and index.
  ISBN 0-8058-2187-2 (cloth : alk. paper). — ISBN
0-8058-2188-0 (pbk.)
   1. Computers. 2. Expert systems (Computer science)
3. Artificial intelligence. I. Hoschka, Peter.
QA76.5.C6142 1996
006.3'3—dc20   95-53124
                                 CIP

Publisher's Note

The publisher has gone to great lengths to ensure the quality of this reprint
but points out that some imperfections in the original may be apparent.

Table of Contents

List of Contributors

Thomas Berlage

Dieter Bolz

Thomas F. Gordon

Wolfgang Gräther

Gernoth Grunst

Joachim Hertzberg

Elke Hinrichs

Peter Hoschka

Klaus Kansy

Werner Karbach

Willi Klösgen

Rüdiger Kolb

Thomas Kreifelts

Peter Mambrey

Bernd S. Müller

Reinhard Oppermann

Franco di Primio

Wolfgang Prinz

Harald Reiterer

Erich Rome

Jörg W. Schaaf

Günther Schmitgen

Markus Sohlenkamp

Michael Spenke

Michael Sprenger

August Tepper

Christoph G. Thomas

Angi Voß

Gerd Woetzel

Stefan Wrobel

Address:

GMD – German National Research Center for Information Technology
Institute for Applied Information Technology (FIT)
D–53 754 Sankt Augustin
GERMANY

1 Computers as Assistants— Introduction and Overview

Peter Hoschka

"The next major metaphor shift in computing will be toward programs that act like assistants rather than tools: they will show more initiative, assume responsibility for larger subtasks, and take appropriate risks." This is the summary of a Microsoft Research project of summer 1994. In a report of the American Association for Artificial Intelligence (AAAI) for the U.S. Congress of August 1994, the presidents of AAAI, Barbara Grosz and Randall Davis, identified "Intelligent Associates: Smart Collaborators" as one of four grand challenges to future research: "To realize them, a common set of capabilities for intelligent systems need to be developed, like... reasoning about the task being performed by each system..., learning from previous experience and adapting behavior accordingly, collaborating..." In July 1994 the *Communications of the ACM* devoted a special issue to "Intelligent Agents," which discusses the basic ideas, views, and insights in this field of research.

Obviously, the possibility of building computer support systems capable of assistance is a compelling one for computer scientists. At GMD, the National Research Center for Computer Science in Germany, a large-scale research program on assisting computers (AC) was begun as early as 1988. The goals of the program were to study and develop the principles and methods for assisting systems, and to demonstrate assistance properties and capabilities in various prototype systems.

Research on computers as assistants means looking for new ways of dividing the labor between humans and computers. On the one hand, future systems should take on more tasks than existing systems do, especially those that seem tedious or difficult for humans. On the other hand, they should *not* automate tasks completely. The basic paradigm is that of assistance. In many fields of application the problems are either too complex or simply too numerous for any attempt to develop a machine with complete problem-solving competence to succceed. What is called for instead is a set of calibrated tools that the user can combine, adapt, and employ as he or she sees fit. Exhaustive treatment and coverage of a problem is in fact *not* the goal of computers as assistants.

There are many characteristics of effective assistance. A human assistant, for instance, is expected to be competent in his or her domain of expertise, to know his or her limitations, to be able to process inexact instructions, to adjust to a client and to learn from the client, and to be able to explain his or her behavior and suggestions. The facilitation of communication and cooperation is a central function of an assistant.

If computer systems are to offer assistant capabilities, they must be supplied with domain knowledge and knowledge about the user. There is an additional requirement: Computer systems need knowledge about themselves, that is, they need to know about the way they function. Only if a system can observe its own behavior, and reflect on it, will it be able to correctly evaluate its own competence and explain its behavior.

The AC concept does not entail building a duplicate of a human assistant. What we tried to do in the AC program was replicate some of the functions of good assistance in computer systems—with no claims to cognitive adequacy. The most important assistant properties we studied in the AC program were the following (Fig. 1.1):

- Domain competence: Assisting computers should be equipped with domain knowledge in certain areas of importance to their users; they should be able to support problem-solving processes in these areas.

- Competence assessment: Within their domain, assisting computers should be able to assess their own competence and their limitations. The user should be able to engage in a dialogue with the system to find out which problems it can solve, which not, and why not.

- Learning and adaptive behavior: Assisting computers should be able to adapt both behavior and functions to a user's individual needs and personal style. The system should learn from the user by monitoring and analyzing his work.

- Processing imprecise instructions: Assisting computers should be able to interpret incomplete, vague, ambiguous, and even contradictory instructions on the basis of knowledge about the user and the current task.

- Explaining abilities: Assisting computers should be able to explain and give reasons for each of their actions, conclusions, and suggestions—in terms the user can understand.

- Cooperation support: Assisting computers exist not only to support the isolated work of individuals, but also the work in teams and organizations. They should help coordinate tasks of a group and provide the organizational knowledge required for cooperation and coordination.

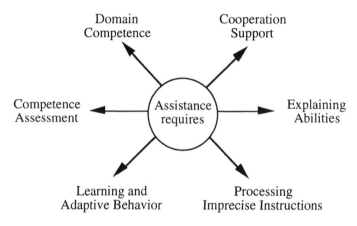

FIG. 1.1 Fundamental assistance properties.

The following chapters describe the concepts and methods we have used in our endeavor to equip systems with assistance properties and capabilities, and the results we attained. This introductory chapter gives a short overview.

COMPETENCE ASSESSMENT

According to the basic paradigm of computers as assistants, we do not try to build systems with exhaustive and perfect competence. But if systems are not perfect, they should be able to assess their own competence. Competence assessment means judging the potential ability of a system to solve a particular problem. More specifically, competent behavior means:

- Before jumping into the solution process, a competent system checks whether it understands the problem, which might be incomplete or ambiguous.

- It does not try to tackle problems that are unsolvable in principle, for instance, if the problem statement is in itself contradictory.

- Nor does it try to solve problems surpassing its capabilities and resources. If necessary, it is able to negotiate the problem statement.

- A competent system is able to detect and remove redundancies and can therefore reduce complexity.

- It monitors its progress and changes focus by adapting strategies or redistributing resources in case progress is behind expectations.

- Finally, it evaluates its solutions in retrospect, so as to store its problem-solving experience. Later, when confronted with a similar case, decisions can be made based on this experience.

Competence assessment requires stepping back and viewing the system from the outside in order to detect its malfunctions. Hence, competence assessment may be regarded as reflective behavior—the inspection of the system by itself. In order to represent the system itself, a reflective system has a meta-level architecture; that is, the part of the system that is reasoned about is represented in a more abstract model at the meta-level (the self-representation). Reflective components inspect, observe, and possibly correct their "object" system from a higher, meta-level. They do not operate on the actual implementation of the object system but on an abstract model of the underlying system. They are thus generic and applicable to a broader class of object systems.

According to this scheme, we built several components for the class of assignment and constraint satisfaction problem solvers. They can recognize overspecified and overcomplex problems, cope with redundancies and contradictions, decompose complex problems, schedule problem-solving steps with respect to time limitations, and switch to a propose-and-revise strategy in case constraint satisfaction fails. A detailed discussion of these problems and solutions can be found in chapter 2, section 2.1.

ADAPTIVE AND ADAPTABLE SYSTEMS

An important requirement for assisting systems is the ability to adapt to users' individual styles and tasks. Two forms of adaptation should be distinguished here: one that occurs on the user's initiative (adaptability), and the other initiated by the system itself (adaptivity or, better, auto-adaptation). Although the call for adaptability is uncontroversial in principle, there are differing positions on auto-adaptation of systems. Our own research has shown that in practice there are only few opportunities for auto-adaptive services. Those that have been implemented so far (adaptation of parameter defaults, control of dialog queries, offering of abbreviated commands) are based only on the evaluation of simple frequencies in user behavior.

System adaptivity is especially desirable in the offering of help when the user is having difficulties. Help facilities should be adjusted according to the current dialogue situation and the individual user. We have developed such a context-sensitive help system for the spreadsheet program Excel. Our HyPLAN system consists of two modules, a plan-recognition program and an interactive multimedia help environment. The plan recognition unit gets a continuous input protocol, listing the commands entered by the user. Guided by a knowledge base of hierarchical action nets, dynamic state models of the user's probable goals are instantiated and incrementally extended as new protocol data comes in. When the user calls for help, the system selects a context-specific help offering based on the goals currently marked. The offerings themselves are realized as voice commented illustrations and animated scenes.

Empirical studies have shown that the possibilites of user-initiated system adaptation are not exploited nearly as intensively as system designers intend. We concluded that the proper exploitation of adaptability requires special support by the system, and thus we looked into the possiblities of using an auto-adaptive component to point out to the user the adaptability capabilities of a system. With FLEXCEL we have devised some new ways of helping the user to utilize the adaptation features of a system. The example application again was Excel. In the meantime, our solutions were incorporated by Microsoft in the new version of Excel (cf. chapter 2, section 2.2).

PROCESSING INEXACT KNOWLEDGE

Assistance systems should be able to perform meaningful actions even if the instructions or the available pieces of information are imprecise and the action to be executed is not definitely prescribed. In this context, imprecision means:

- The available information can be incomplete, so that decisions must be based on plausible additional assumptions (defaults), which may have to be revised later.

- The available information can be vague or ambiguous, for example, because certain relationships are true only with some probability, or because terms used in an instruction have no precise meaning (draw a triangle *near* the rectangle).

- The available information can be contradictory, which means that parts of it must be faded out and left out of consideration.

As an illustrative field of application, we took the problems of processing imprecision during the construction, editing, and search of graphics. The user can incrementally specify the attributes of graphical objects. The assistance function consists of completing or making more precise and consistent user specifications that are imprecise in the foregoing sense. The methods we used here were mainly concentrated on nonclassical methods of inference, such as nonmonotonic (i.e., based on revisable assumptions) and associative reasoning (cf. chapter 2, section 2.3).

Nonmonotonic inference mechanisms are of interest because they allow the assisting system to complete imprecise problem specifications with the help of standard assumptions. Default rules make it possible to infer plausible conclusions, which can lead to meaningful decisions in the case of incompleteness.

Associative reasoning techniques based on neural networks are used for the storage and retrieval of known problem specifications and solutions. The solution of a current problem is supported by looking at similar, already solved problems. These are reactivated in a content-driven, associative manner (cf. chapter 2, section 2.3 and chapter 3, section 3.5).

EXPLANATION ABILITIES

If assisting systems process imprecise instructions, make suggestions for system adaptation, and offer domain competence for solving problems, then these capabilities inevitably give rise to a further requirement: The systems must be able to explain their own behavior and suggestions—in terms the user can understand. Assisting systems have no chance of being accepted as black boxes. It is well known of existing expert systems that the explanation components actually explain very little. They are usually limited to confronting the user with a more or less flexible presentation, giving the protocol for each of the the problem-solving steps executed. Explanation ability in the assisting computer sense, furthermore, entails not only making a system more transparent, but also implementing pedagogical competence so that the user's understanding of a problem can be estimated and improved in the course of dialogue. Consequently, a system must have tutorial faculties, and not be limited to simply showing the formal structures of a knowledge-based system.

As a first step toward better explanations, we worked on improving knowledge representation methods so as to make them more suited to explanation purposes. We examined the possibilities of using conceptual models that represent knowledge in several levels of abstraction to determine the focus and level of detail for an explanation to be produced (cf. chapter 2, section 2.4).

SYSTEMS WITH DOMAIN COMPETENCE

Supplying systems with domain knowledge is the goal of conventional expert systems. A number of examples that deviate somewhat from the usual expert system scenario show how assisting computers can incorporate some of the knowledge that must, at present, be supplied by the user. The major difference between conventional expert systems and assisting computer systems involves the roles of the human and computer. Most expert systems ask the user for input, make all decisions, and then return an answer. In assisting computer systems, the user is an active agent empowered by the system's domain competence. This means that the systems and the person are bringing complementary strengths and weaknesses to the job. Rather than communicating with computers, users should perceive themselves as communicating with application domains via computers. To shape the computer into a truly usable and useful medium, we must let users work directly on their problems and their tasks.

We have created (chapter 3) five paradigmatic assisting systems with domain knowledge:

- An assistant for knowledge acquisition that supports the development of a knowledge base (section 3.1).
- A statistics interpreter that helps to detect interesting findings in statistical databases (section 3.2).
- A graphics designer that assists the design of business and presentation graphics (section 3.3).
- A design assistant and a graphical search tool that use case-based memory and principles of gestalt theory to support the reuse of graphical objects (sections 3.4 and 3.5).

- A user interface design aid tool that assists the designer of graphical user interfaces during the design process (section 3.6).

All these systems incorporate domain competence. They also, however, contain elements of other assistance properties. The knowledge acquisition system, for instance, is able to learn from the user and identify and manage imprecise instructions. The graphics designer system can also process imprecise instructions: The user develops a graphic to a certain point and then passes on this raw version to the system, which takes over the details of producing an aesthetically pleasing end product.

Assistant for Knowledge Acquisition

A system to support the acquisition of knowledge cannot be limited to offering the user some formalisms for the representation of knowledge. Acquiring knowledge in a new domain is not a straightforward task that can be easily planned out for computer treatment. It is a creative process with frequent mental leaps and many iterations. The system must be flexible so that the user can work in his or her own manner and is not forced to adapt to the system's. The system must be able to supply the user with the consequences of new knowledge that the user adds to the system. Since a domain model is built in the course of work with the system, it must be able to accept contradictory, incomplete, and preliminary inputs. We call this the paradigm of "sloppy modeling" (Morik 1989). Finally, the system should help the user complete and rectify the model, for instance by suggesting new rules and concepts to supplement the existing knowledge. These requirements have guided the conception of the knowledge acquisition system MOBAL (model-based learning system).

MOBAL is a workbench offering several tools for knowledge acquisition that can be used individually or in combination. The heart of the system is the Coordinator & Inference Engine, which manages the knowledge base and coordinates the system's various functions. MOBAL uses facts and rules to represent a domain model. Existing facts and rules can be modified at any time; the inference engine makes sure that entries relying on the changed ones are modified as well. If entries are incomplete or erroneous, they do not have to be corrected immediately. The system maintains an agenda of such "open ends," and the user can address these at any time he or she sees fit. Even contradictory entries can be processed; the system recognizes contradictions and offers a special tool, the Knowledge Revision Tool, with the help of which a contradiction can be analyzed and resolved.

The other tools that comprise MOBAL serve the purpose of supporting the user with various machine learning techniques. They produce suggestions that can help resolve incompleteness, improve knowledge-base structure, or recognize and point out regularities in the facts. The user can use these suggestions or decide to make similar (or other) entries. All machine learning techniques in MOBAL play the role of an assistant to the user. Modeling is understood as a cooperative and balanced process between human and computer.

Statistics Interpreter

A second example of an assistance system with domain competence is the statistics interpreter Explora. It discovers interesting findings in statistical data and aide the user in interpreting them. The system has knowledge about the domain and about methods for the analysis of statistical data. To this end, the objects, relations, and queries used in the statistical analysis of a data set must be represented in the system. From these, the system constructs a proposition space for these data and searches it for interesting findings. In traditional statistics packages, the user must specify each hypothesis individually. In Explora, this is different: Knowledge-based and systematic processing of the proposition search space allows the system to discover results that might have been overlooked in traditional analysis. The statistics interpreter usually has to cover a very large search space of potential findings. However, this space can be pruned substantially with the help of redundancy filters and generalization methods. This assures that only the "strongest" findings are presented to the user and the user is not flooded with redundant statements.

Explora accepts propositions of an arbitrary type which are useful to formulate knowledge about data and incorporates them in a "discovery system." In other words, the system is not limited to specific types of hypotheses (as, for instance, rule learners are). Explora also supports the user in navigating through the space of potentially interesting findings by being able to refine, specialize, condense, and generalize propositions. Explora is an early (and working) system in the area discussed nowadays under the headings of data mining or knowledge discovery in databases.

Graphics Designer

A further example of success in incorporating domain knowledge in a computer system is the graphics designer. Today's graphics packages allow anyone to produce graphics easily and comfortably. Current systems do not, however, offer any assistance where aesthetic beautification or the choice of suitable graphical means of expression is concerned. It is here that the graphics designer aims to provide assistance.

In creating graphics, certain rules of design that govern expressiveness must be observed. These rules state, for instance, which diagram type is suited for time series, and which for rank orders, or which colors make a good background and which a good foreground. The layperson is usually unfamiliar with such rules and sometimes they elude even professionals. We have developed a system that helps the user in producing a suitable business graphic.

In the fine calibration of a picture, graphical elements must be aligned to each other and spread out evenly across the available space. The more precisely a graphic can be printed, the more conspicuous discrepancies in layout and slight inaccuracies in size and position become. The graphics designer can describe typical errors in graphics in a "situation language" we have developed, and can automatically find these errors. A knowledge-based criticism module decides which corrections are to be made in which order, and plans out the individual steps in such a way as to ensure that later corrections do not reverse previous ones.

Graphical Search Assistance

More and more products are designed with graphical editors, such as simple drawing tools, CAD (computer-aided design) tools, or graphical programming language editors, and are then archived electronically. The logical next step is to reuse stored designs for other products, either as a source of inspiration, or directly via copy, paste, and adaptation. Retrieval of archived designs is a problem if there is no foolproof filing scheme and if one does not exactly know what to look for. One may be forgetful, designs may have been produced by other persons, or designs from different contexts can be mistakenly appropriated.

Here, the most suitable support would be an assistant that can semantically interpret graphical queries and perform a so-called content-based retrieval. Such an assistant could be given a sketch and asked to retrieve something similar and hopefully more complete. One challenge in interpreting such fuzzy queries is to detect structures and forms that were not explicitly drawn, so-called emerging shapes. Techniques for that purpose often rely on principles of gestalt psychology, such as the one presented in chapter 3, section

3.4. Another challenge is to find a fuzzy notion of similarity, which typically is domain dependent. Two such techniques are discussed in chapter 3, section 3.5. All approaches have in common that design documents need not be indexed manually, because the same mechanism that automatically interprets a query can also interpret the designs archived. Thus, they free the designer from boring work in order to concentrate on the more creative aspects.

User Interface Design Assistant

In the effort to provide high ergonomic quality graphical user interfaces, designers have acquired knowledge about human factors and expressed it in standards, style guides, and guidelines. The volume of these available sources is already huge. Simultaneously, user interface designers need more and more competence, knowledge, and experience to handle this great amount of human factors knowledge. This means for most of them that optimally performing their jobs requires taking into account far more information than they can possibly keep in mind or apply. This results in a need for user interface development tools with domain competence based on human factors knowledge that may be encountered, learned about, practiced, and improved during ongoing use—in other words, tools with which users learn and use on demand. An important research goal in the development of user interface tools is, thus, to discover helpful, unobtrusive, structured, and organized ways to integrate human factors knowledge into the tools without stifling creativity—and to provide the designer with assistance in understanding, searching, and applying this knowledge. This goal was the starting point for the project IDA (User Interface Design Assistant). The primary purpose was to explicitly incorporate domain competence in user interface development tools to empower the user interface designers. The following benefits for user interface designers using the IDA development tools are expected:

- Designers will be able to acquire human factors knowledge during their daily work using their development tools ("learning and using on demand").
- Designers will be enabled to apply ergonomic style guides and guidelines ("usability").
- Designers will be able to use predefined ergonomic user interface components ("reusability").
- Designers will be able to evaluate the ergonomic quality of their design during the design process ("quality assurance").

SUPPORT OF COOPERATION

Every form of activity in organizations requires cooperation among its members. Assisting computers should therefore not only support the isolated work of the individual but also help coordinate activities within and across groups, and provide access to the necessary knowledge about the organization. In the AC program, these functions of assisting computers were analyzed and prototypically implemented in the following components (chapter 4):

- A task manager that helps coordinating group work (section 4.1).
- A mediating system for the support of discussion, argumentation, and decision-making in groups (section 4.2).
- An organizational knowledge assistant that offers information about structures and procedures of organizations (section 4.3).
- Experiments with new metaphors to effectively present complex cooperation support environments (section 4.4).

Coordination Support by Task Management

As a contribution to the effort to provide more effective computer assistance for cooperative activity, we have developed the Task Manager, which is especially oriented to supply coordination support for geographically distributed work situations typical of large business organizations and government agencies.

With the help of the Task Manager, users may organize cooperative tasks, monitor their progress, share documents and services, and exchange informal notes while performing their tasks. The Task Manager distributes task specifications, attached documents, and notes to the involved users in a consistent way. It is meant to support the management of work distributed in time and/or space by providing:

- Support of organization and planning of collaborative work (who does what, with whom, until when, using what material).
- An up-to-date overview of collaborative activity and work progress.
- Dynamic modification of work plans during performance.
- Availability and exchange of documents and informal notes within groups of people involved in task performance.

Mediating System for Argumentation and Decision Making

Another cooperation support system that we developed is a "mediating system" for supporting discussion, argumentation, and decision making in groups over electronic networks, the Zeno system. The system is designed to take resource limitations and conflicts of opinion and interest into account. It is a contribution to what might be called "computational dialectics," the study of computational models of norms for rational discourse.

The thesis is that rationality can best be understood as theory construction regulated by discourse norms. Techniques from artificial intelligence are applied for supporting practical reasoning given such pragmatic constraints as limited time and uncertain, incomplete, or even inconsistent information. To avoid the rigidity inherent in formal models of norms and to facilitate an acceptable balance between the individual interests of autonomy and management interests of control and accountability, the space of possible actions is kept cleanly separate from the space of permissions and obligations. That is, our mediating system advises participants in a discussion about their rights and obligations, but does not automatically enforce rules. Using the legal system as a model, Zeno distinguishes between the legislative and judicial functions. Participants have an opportunity to argue about whether some rule should be applicable in a concrete case; the norms evolve and adapt through use.

Organizational Knowledge Assistant

Cooperation in teams and organizations is embedded in an organizational framework. Thus, information about the organizational context in which users work helps to choose the right patterns for communication and cooperation. Information must be provided to answer questions such as: Who is responsible for carrying out a specific task? Whom can I ask for help? Furthermore, the system should provide information as to how particular tasks are handled in the organization. What are the organizational rules one has to consider? Whom do I have to ask first? Which document type do I have to use? All this information is part of the knowledge that is normally not or only very implicitly provided by cooperation supporting applications, although it plays a significant role in cooperation. We developed a system, named TOSCA, that provides this information to users and applications.

The organizational knowledge assistant aims at two goals. The first goal is providing the organizational knowledge required to ease the integration of cooperative applications in an organization. In the service of this goal, it offers well-defined interfaces and supporting distributed management. In combination with a user interface it is meant to cater to the informational needs of the end user.

The dynamic nature of any organization makes it impossible to develop a single representation that fits all organizations. Thus the second goal is ensuring "tailorability" that allows an adaptation of the system to various organizational settings. TOSCA heightens the visibility of the relevant concepts and provides an object modeling tool that allows users and groups to tailor the object model to their specific needs.

Dynamic Interfaces for Cooperative Activities

The success of computer-supported cooperative work depends to a large part on the quality of the user interface. User interfaces for cooperative applications are demanding because the environment is no longer static and the user has to be aware of other people's actions. This is a significant departure from the single-user desktop metaphor. New metaphors as well as innovative techniques (such as animation) are required to effectively present the complex working environment on a small computer screen.

The DIVA (Dynamic interfaces for cooperative activities) project has identified the following three key problems to improve the interfaces for cooperative applications:

- A new metaphor is needed for cooperative work. The conventional desktop metaphor is no longer sufficient, because the private desktop must be expanded to a space of cooperating people.
- The concept of time becomes more important. In particular, users must be supported by a history of activities because it is not possible to memorize other people's actions.
- Cooperative activity must be visualized. New mechanisms, including three-dimensional structures and animation techniques, are required to provide the level of awareness necessary for successful telecooperation.

We have developed a prototype that tries to mimic on the screen the social behavior in a real office, to facilitate communication between users and the use of cooperative applications. We experimented with a first version of the "virtual office" metaphor that represents the environment as people moving between rooms in a virtual building.

METHODOLOGICAL AND TOOL PROJECTS

The overall AC effort was planned to consist of many components—the assistants—with consistent user interfaces designed according to uniform guidelines. Therefore, a common development base for all projects was defined: the UNIX operating system, the X Window System, the OSF/Motif graphical user interface, and LISP/CLOS (Common LISP Object System) and C++ as programming languages. However, it soon became clear that this decision alone was not sufficient to guarantee consistency, and therefore the project GINA was created in order to define guidelines and to incorporate them into a software tool to be used by those involved in the projects implementing assistants. In chapter 5, section 5.1 we give an overview of GINA.

In the context of the AC program there were several additional projects that focused not on the development of specific user-oriented functionality but more on the solution of basic methodological problems. Thus, we intensively studied the potential of artificial intelligence planning for building assistance systems. Planning can be both a purpose and a means in assistance systems: A purpose whenever the task requiring assistance is a genuine planning task, and a means whenever an assistance task or property requires some internal planning. Section 5.2 contains the results of our work on planning in this context. In section 5.3 we investigate the instrumental use of guiding visions and metaphors, such as the assistance metaphor, for designing computer systems and for preventively oriented technology assessment.

CONCLUSION

Work guided by the concept of computers as assistants began in 1988. The results reported in this book show that we have come closer to our goals and have been able to demonstrate, at least in sample fashion, some of the assistance capabilities we aimed at. New methodological approaches to a number of problems have been found. But for the most part, we are still in the early stages of progress toward our goals. Therefore, we are pleased that so many other research groups have also started work on computers as assistants.

ACKNOWLEDGMENT

A large number of people were involved, directly or indirectly, with our AC research program on assisting computers. Above all I would like to thank all the members of the AC program for their support and constructive work. Particularly the project leaders Franco di Primio, Klaus Kansy, Willy

Klösgen, Thomas Kreifelts, Katharina Morik, Reinhard Oppermann, Horst Santo, Michael Spenke, August Tepper, and Angi Voß always supported the basic idea of building computers as assistants. I would like to thank my colleagues Thomas Christaller and Peter Wißkirchen for their support in directing the AC program; particularly Peter Wißkirchen always was there with help and advice when problems and impediments arose. Last but not least, I am extremely grateful to Klaus Kansy for his invaluable assistance in the editorial work for this book. Erich Rome has put a lot of effort into polishing the references. Gabi Vezzari patiently kept the editorial process running and helped with the subject and author indices. I gratefully acknowledge the help of Lawrence Erlbaum Associates, particularly of Susan Milmoe, during the publishing process.

2 Systems with Assistance Capabilities

2.1 COMPETENCE ASSESSMENT

Angi Voß, Werner Karbach

An assistant, we expect, should free us from repetitive and routine work. We would like to delegate any well-understood tasks to the assistant. But delegation must be worthwhile. If it takes too long to explain the task or to provide the relevant information, or if our assistant gets stuck too often or takes too much time, we feel we had better do the job ourselves. In short, assistants should not only be experts in their domain, but they should handle their tasks competently.

The REFLECT project[1] was aimed at exploring how knowledged-based systems as experts could be turned into competent problem solvers. Like any assistant, a competent system need not be perfect. But it should not try to solve unsolvable problems. Rather, it should simplify and decompose over-complex problems, complement missing information, weigh the quality of a solution against the cost of achieving it, approximate solutions if time is sparse, and so forth.

To make a knowledge-based system more competent, we proposed to enhance it by suitable competence specialists. By this we meant independent generic modules, each devoted to a special type of competence improvement.

[1] The research project REFLECT was partly funded by the ESPRIT Basic Research Programme of the Commission of the European Communities under no. 3178. The partners for this project were the University of Amsterdam (The Netherlands), the GMD (Germany), the Dutch Energy Research Foundation ECN (The Netherlands), and BSR-Consulting (Germany). This chapter originates from the research project REFLECT and is based on the article "Building competent reflective systems", in A. Yonezawa and B. Smith, (Eds.); Proceedings of the International Workshop on Reflection and Meta-level Architecture, Tokyo, 1992. A German version appeared in Der GMD-Spiegel 2, 1992.

The resulting system is a reflective system because it reasons about and modifies its own problem-solving behavior. We developed a framework to describe such systems at the knowledge level and implement them afterward in a structure-preserving way. We applied this framework to improve a knowledge-based system for assignment problems by 10 competence specialists.

Our Problem Definition

Knowledge-based systems are used to solve difficult problems. But they often fail abruptly because they had no idea of their own abilities. We wanted to lay foundations for future knowledge-based systems to know more about the limits of their own competence and to use this knowledge flexibly. In such systems, so was our hypothesis, reflection plays a central role. The present paper is to show the outcome of this hypothesis.

What Is Competence?

Competence does not mean that one knows everything or that one makes no errors. Competent systems, however, know what they do not know and they recognize errors and deadlocks in their approach as quickly as possible (Voß et al., 1990). They can:

- Estimate whether a problem is solvable.
- Modify a problem to make it solvable.
- Assess the effect of missing or uncertain information.
- Simplify problems before searching for solution.
- Estimate the cost of solution finding.
- Compare different strategies.
- Recognize deadlocks and search for ways out.
- Classify problems and compare them with each other.
- Request support from outside.
- Handle time restrictions.

This list shows that, in addition to pure problem solving, there are mechanisms that are used not for actual solution finding but to monitor one's own approach from a higher level, and that intervene if required. A competent system will reject unsolvable problems, for example, or it will reduce them to make them solvable.

The integration of meta-knowledge into knowledge-based systems has already been discussed very early in AI (Davis & Buchanan, 1977; Davis, 1976; Sussman, 1975). Typical meta-activities considered were strategic control (Hudlicka & Lesser, 1984) and meta-planing (Stefik, 1981; Wilensky, 1980), recognition of deadlocks in problem solving (Jansweijer, 1988; Rosenbloom, et al., 1988; Sussman, 1975;), explanation of knowledge bases (Davis, 1976), and acting under time restrictions (Russell & Zilberstein, 1991). In logic, too, meta-levels have been studied (see (van Harmelen, 1991) for an overview), in particular with respect to self-referential phrases (Perlis, 1985).

The Object System

REFLECT discussed two applications, an analytical application and a synthetic one. Our Dutch partners were concerned with diagnosis on the basis of error models and with the aid of qualitative simulation, while we studied an assignment problem solver (Bartsch-Spörl et al., 1991). Its domain-specific instantiation is called OFFICE-PLAN because it is used specifically for the assignment of employees to offices. We next explain the task and functioning of OFFICE-PLAN briefly and then show the incompetence of the system and what to do to remedy it.

Figure 2.1 sketches the object system in a very generic terminology.

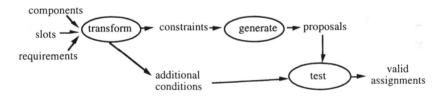

FIG. 2.1 The three inference steps of our assignment system OFFICE-PLAN.

Detecting Causes of Incompetence

Although OFFICE-PLAN always provides all correct solutions, it is fairly incompetent. Mostly it needs a lot of time for solving difficult problems without any warning so that the user could reduce his or her requirements. It cannot find approximate solutions to inconsistent problems, although realistic problems tend to be overspecified.

Within REFLECT we have found out that incompetent behavior can often be explained by the following causes, which may occur both in the problem definition and in the permanent knowledge base (Voß et al., 1991):

Complexity: In OFFICE-PLAN, a problem is often complex if the number of employees and the number of available offices are almost equal or if there are many restricting constraints and conditions.

Inconsistency: In OFFICE-PLAN, inconsistent requirements and conditions may be in the problem definition.

Irrelevance: There may occur various redundancies between constraints and conditions in OFFICE-PLAN.

Underspecification: This occurs, admittedly rarely in office planning, where the constraints and conditions restrict the possible allocations too little.

Overspecification: In OFFICE-PLAN, it is almost normal that not all constraints and conditions can be fulfilled simultaneously.

Uncertainty: In the REFLECT applications, we have no problems in this respect because all information is categorical.

Errors: In the qualitative simulation system of our Dutch colleagues, an error occurs whose elimination would require the scanning of a gigantic Prolog program.

Competence Improvement as Reflective Behavior

How can a given system recognize and possibly remedy its incompetence in a given problem setting? We suggest a separate instance that observes the given incompetent system, examines it for competence weaknesses, and intervenes correspondingly to enable the object system to continue its work more properly.

Consequently, in addition to the original problem solver, we get a second one at meta-level whose subject is the original problem solver. Such interacting combinations of object system and meta-system are referred to as reflective systems in the literature (Maes, 1988). Indeed, an external observer would attribute to the overall system a behavior that we usually call reflective: "It is stuck, it reflects on the problem. It identifies a contradiction and makes a compromise and now it continues computing."

Results from cognition research make us assume that people, too, interrupt their reasoning to reconsider their problem at a higher level (Brown, 1984).

According to Flavell (1984), this requires both knowledge about the abilities of the own cognitive apparatus and knowledge about the monitoring of problem solving steps. Kluwe and Schiebler (1984) subdivided the latter into the identification of the task, the examination and evaluation of the current state of problem solution, a prospect, and the focusing on important subproblems. During experiments, one observed that probands did better if they were encouraged to reflect on their own approach (Dörner 1989).

Accordingly, we consider competence assessment and competence improvement by a meta-system as somewhat like process control, which consists of the generic tasks dynamic analysis, interpretation, diagnosis, and repair of the processes running in the object system if required. Figure 2.2 shows the typical inference steps.

Because it is now a problem solver on its own, for accomplishing its monitoring function the meta-system must of course possess meta-knowledge about the object system, its task, the application domain, and general knowledge about good and bad problem-solving behavior.

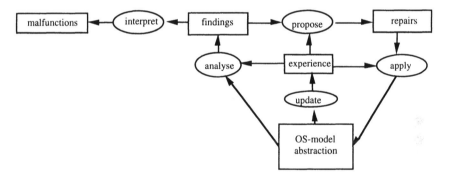

FIG. 2.2 Inference steps of the meta system for competence monitoring.

Generic Specialists for Competence Improvement

To cope with the various incompetences of constraint-satisfaction-based assignment problem solvers like OFFICE-PLAN, we have devised a total of 10 different modules, each oriented to a specific type of incompetence. All modules have been designed and implemented as meta-components of a reflective system. In the sense that OFFICE-PLAN is a generic assignment problem solver, each of our 10 modules itself is a generic competence improver. When presenting the modules briefly in this subsection, we discuss in detail these general aspects. We describe the assumptions about the object systems the individual modules are based on:

FEASIBILITY examines the feasibility of the transformed problem by comparing the available resources with the required ones. If there are not enough rooms to accommodate all employees, for example, the problem will be unsolvable. If there are roughly enough, it is probably difficult to solve. FEASIBILITY can be applied to all problem solvers that have to observe certain resource restrictions.

CONTRADICTORY-R and CONTRADICTORY-C recognize pairs of inconsistent requirements and/or constraints and enable the user to withdraw one of them. CONTRADICTORY-C, for example, would recognize that two employees must be placed into the same office or into different offices because they must cooperate, or because only one of them is a smoker. CONTRADICTORY-C can be used for any constraint problem solvers, and CONTRADICTORY-R only for those that include a step transforming requirements into constraints.

REDUNDANT-R, -C, -CF and SIMPLIFY-DOMAIN simplify the problem or its transformed version. REDUNDANT-R removes redundant requirements; REDUNDANT-C and REDUNDANT-CF remove redundancies between constraints or between constraints and conditions that, for example, make two employees sit in different offices or in the same office. Their range of use corresponds to that of CONTRA-DICTORY-R and CONTRADICTORY-C. SIMPLIFY-DOMAIN eliminates constraints that are covered by a condition. For example, two employees do not need to sit in different offices if one does not want to share his or her room with a colleague. This module is reusable in problem solvers that distinguish between the generation of constraints and the testing of conditions.

DECOMPOSE&RELAX decomposes difficult problems to subproblems, which it submits to OFFICE-PLAN step by step. It does so only if it has not found any solution of the subproblem in its case collection. If a subproblem is not solvable, it decomposes it into further constituents and tries to solve them separately as described. In this way, it approaches compromises in the case of unsolvable problems. DECOM-POSE&RELAX is very fast because not only the constraints but also the conditions are tested for each subproblem, which limits the search space dramatically. In addition, DECOMPOSE&RELAX does better when its subproblem collection is larger. It can be based on any constraint systems; however, it is more effective if these have separate generation phases and test phases.

TACKLE-TIME tries to find a given number of solutions within a given time interval. For this purpose, it intervenes into the generation and test phase of the object system. It attempts to generate as few as possible allocations so that, after testing, just as many allocations are left as the user has specified. TACKLE-TIME is usable in any system working with the generate and test method.

CASE-BASED has a library of solved, unsolved, or relaxed cases. Given a concrete problem, it can find out fairly fast whether it knows this case and therefore its result. Otherwise, it cannot process the problem. CASE-BASED assumes that a problem is not solvable if one of its subproblems is not solvable. Whenever this condition is fulfilled, one can use this specialist.

Model-based competence improvement

When describing our competence specialists, we have made explicit the knowledge employed and the specialists' assumptions about the object system. The assumptions are very abstract and totally implementation independent. This means, instead of applying our modules directly to the object system, we can also base them on an abstract and partial description of that object system. The advantage would be that we could reuse such a module directly without any modification for another object system. This idea distinguishes our approach from reflective programming languages such as 3-Lisp (Smith, 1982) or KRS (Maes, 1987), which are based on implementational aspects such as the interpreter. We can reuse FEASIBILITY, REDUNDANT-C, or CONTRADICTORY-C in order to improve an assignment problem solver using a hill climbing approach, for example. Such a system would place one employee after the other into the offices by always selecting the most promising move. Using this method, we get approximate solutions for inconsistent problems, but possibly only suboptimal solutions for consistent problems. In more pictorial terms, the system can get stuck at a local maximum (hill).

Everything has its price, and so has the generality we achieve. As the competence specialists are based on an abstract description of their object system, we must ensure that this model is consistent with the object system, at least in the moments where the systems synchronize. In technical terminology, one speaks of a causal connection. Whenever the meta-component inspects its model, it must reflect the object system. And whenever the meta-component modifies the model, this intervention must be propagated into the object system. In REFLECT, we have elaborated this very approach and validated it on the basis of our applications (Schreiber et al., 1991).

Knowledge Level Models

Because we want to base our competence specialists on an abstract model instead of basing them directly on the object system, we have to clarify what such a model should look like. There are indeed not many approaches to the modeling or specification of knowledge-based systems. The most concrete frameworks are Chandrasekaran's generic tasks (Chandrasekaran, 1988), Steel's components of expertise (Steels, 1990), the role-limiting-method approach (Klinker et al., 1990) and the KADS[2] models of expertise (Breuker & Wielinga, 1989) (see (Karbach et al., 1990) for a comparison). They all claim to be knowledge level descriptions of generic tasks that appear repeatedly in knowledge-based systems, like diagnosis, assessment, prediction, or configuration. The term "knowledge level" was created by Newell (1982).

We have decided to use the KADS models because they provide the finest differentiation. They distinguish domain knowledge, inference knowledge, and task knowledge. In the case of OFFICE-PLAN, the domain is the office world, the inferences are mainly those shown by Fig. 2.3 and the task knowledge defines the order of executing these inference steps. For example, all candidates could be generated before testing them together, or the candidates could be generated and tested stepwise. In principle, it might even not be necessary to generate or test all candidates. The KADS methodology suggest designing such models during knowledge acquisition and using them as a communication medium. They exist on paper only and can be restructured as desired in the course of implementation. Therefore, the complete system in general need not reflect the model.

For our approach in REFLECT, it is significant that the models are represented in a formal language so that the computer can interpret them. In addition, they must correspond to the operational system as directly as possible. It would be desirable to couple model and system in such a way that a change to the system is directly reflected in the model and vice versa. Ideally, the

---

[2] One reading of KADS is Knowledge Acquisition and Document Structuring, another one is Knowledge Acquisition and Design Structuring.

model should be a constituent of the system or, in other words, the system should be an extension of the model around all operational details. For this purpose we have developed a language called MODEL-K (Karbach &Voß, 1992b).

As shown in Fig. 2.3, we even went one step further. Because the meta-component, too, is a knowledge-based problem solver, the same modeling technique can also be used at meta level. In this way we are able to describe our competence specialists not only independently of many details of the object system but also independently of any detail of their own implementation. The extension of MODEL-K to the description of meta-systems is uncritical—except for one aspect concerning the coordination of our competence specialists. So far, we have always assumed that we describe a reflective system with one object system and one meta system and not with a whole set of meta-systems.

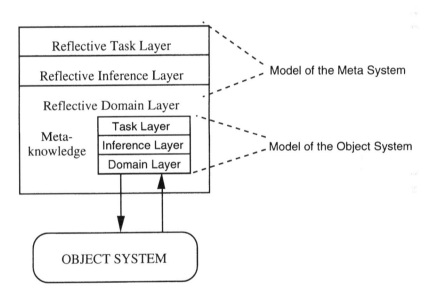

FIG. 2.3 Framework for descriptions of reflective systems.

Integration of Generic Competence Specialists

Indeed, we developed the individual competence specialists independently of each other, in such a way that each competence specialist, together with the object system, formed a fully operable reflective system. In the course of time, we got modules that could be combined reasonably. A first configuration of such types were the problem reformulation specialists REDUNDANT-..., CONTRADICTORY-..., and SIMPLIFY-DOMAIN. Later we combined DECOMPOSE&RELAX with FEASIBILITY, the latter with TACKLE-TIME, and finally we integrated them all. Because this approach was chosen very early, we searched very consciously for a suitable development method. When combining several specialists, we must modify their (one-sided) coordination with the object system on the one hand, and, on the other, we must coordinate the specialists among themselves. For that purpose, we decided to shift all coordination aspects to a separate top level. For a different combination of the modules, we need only exchange this coordination level; the rest remains unchanged.

The functionality of the original system has been extended by the reflective extension. Due to DECOMPOSE&RELAX, approximate solutions for inconsistent problems are now possible; thanks to TACKLE-TIME, we can determine in advance the time and number of the solutions to be obtained; and CASE-BASED can possibly lead to a very fast solution. The system has become more versatile and more useful and it is easier to adapt it to the current needs of its user. The competence specialists provided three new degrees of freedom in system use (Voß et al., 1992). We may admit compromises or not, we can limit time precisely, and we can influence the number of solutions. Figure 2.4 shows how we dynamically configure the specialists depending on the user's need, and how we coordinate them with the object system.

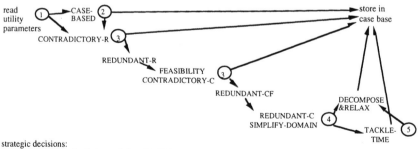

strategic decisions:
1: if time = quick and solutions=all then CASE-BASED
2: if no case found then CONTRADICTORY-R
3: if consistent or relaxation=yes then REDUNDANT-R
4: if consistent and time=bounded then TACKLE-TIME else DECOMPOSE&RELAX
5: if inconsistent and relaxation=yes then DECOMPOSE&RELAX

FIG. 2.4 Coordination of competence specialists.

The alternative, to activate OFFICE-PLAN alone, is not considered anymore. Our experiments have shown that it always does better in combination with the competence specialists. Figure 2.5 gives an idea of the behavioral changes caused by the reflective modules. The evaluation criterion is the number of generated solutions. As compared with OFFICE-PLAN alone, the redundancy-eliminating modules reduce the necessary time considerably. Nevertheless, the user must wait for the generation of all applicable solutions even if the user is interested in only few ones. Only the use of TACKLE-TIME causes a gentler transition during the generation of solutions. By using different strategies at meta-level, we can generate different performance profiles (Coulon et al., 1992).

FIG. 2.5 Performance profiles of OFFICE-PLAN alone with redundancy-eliminating modules and with additional time control measured by the number of generated solutions.

Evaluation

At the beginning of REFLECT, we thought that specific functionalities such as competence assessment, but also explanation and support, should be reflective in principle. Judged by the insight since then, this assumption was incorrect. Any behavior specified at meta-level can be "compiled down" to the object system. The approach developed by us is therefore not imperative, but it has some advantages.

Conceptual adequacy: Behavior to be regarded as meta-activity in conceptual terms should also be described as such.

Modularity: The description at the meta-level is modular and can be far better understood than by a direct extension of the object system.

Reusability: By basing the meta-components on abstract descriptions of the object system, the knowledge model, they become immediately reusable for all those object systems that correspond to this model.

Configurability: By supporting the configuration of several meta-modules, we increase their reusability even further. By gathering a lot of meta-modules, we should be able eventually to enhance any system by simply selecting the suitable meta-modules from a library and by coordinating them appropriately.

Implementation independence: One may argue that the implementation of a meta-system is less efficient than the direct extension of the object system; or the competence specialists may not executable in the new target environment. However, this is no problem. By first describing the competence specialists in abstract terms, at least these descriptions will be reusable as implementation patterns.

For the reasons just mentioned, we recognized in the course of the project that our approach provides a potential as an evolutionary software development method going beyond the actual project objectives.

Acknowledgment

We thank Brigitte Bartsch-Spörl for her comments on this section, and we thank Carl-Helmut Coulon and Uwe Drouven, who developed and implemented TACKLE-TIME and DECOMPOSE&RELAX, respectively. Bernd Linowski helped to transform the document from Latex to Word.

2.2 ADAPTIVE AND ADAPTABLE SYSTEMS

Gernoth Grunst, Reinhard Oppermann, Christoph G. Thomas

This section discusses the ways in which assisting systems can be made adaptable or adaptive[3]. System adaptation is mainly used to bridge the distance between development and application. One of the basic problems involved in the development of application software is to identify typical situations of usage for the product to be developed, such that the product can support the activity of the target group in a "task-specific" and "user-specific" way. This will be more successful when the interaction of software development and applicational field is closer, the definition of the activities to be supported is more exact, and the special work styles of the people concerned are better known. However, application systems are not developed for an individual user or an individual task. They should rather be widely usable: for many users, for a task spectrum, over a longer period. This applies even to special developments for a specific user, and even more to widespread standard software. This increases in general the distance between development and use.

This distance is bridged by an increasing complexity of systems and an orientation to the average user to be able to operate an application range as wide as possible by means of a great number of available functions. This solution is not satisfactory. The average user and the average task do simply not exist. Situations exist in which the system does not satisfy the actual requirements. It does not provide the required function, its operation is too complicated for specific recurring work flows, the name or placement of functions or objects does not correspond to the user's ideas, and so forth.

One solution to these problems is to extent the development phase to the utilization phase. This should reduce the distance between development and application in a form to be mastered by users. However, the possibilities of system maintenance and adaptation by the developer are limited for economic and organisational reasons—all the more so in the case of standard software. Therefore there are attempts to provide the end user with facilities and tools for adapting the system to the tasks to be handled and to specific

---

[3] This section was written for the SAGA project, supported by the German Federal Ministry for Research and Technology from December 1988 to November 1991 in the framework of the program "Work and Technology."

habits and requirements. These facilities not only relieve the developers from the requirement of individual system design, but they also consider the differentiation requirements of ergonomics.

In particular, human engineering findings concerning the personality-supporting effect of freedom in working have led to the demand that information technology systems should prescribe not only a specific type of information use and processing (which the system developer considers to be correct), but that they should also enable the development of individual strategies of problem solving from the user's point of view. To develop work systems that follow a "one best way" of all conceivable work forms is considered questionable in general (Ackermann & Ulich, 1987, pp. 131ff.; Grob, 1985; Ulich, 1978; Zülch & Starringer, 1984). The user should be supported to discover or to create action spaces, to influence traditional forms of task accomplishment to change, and to adapt them to new situations in a dynamic manner.

Adaptivity supports the general ergonomic characteristic of flexibility. Flexibility may consist of a variety of available interaction techniques provided in parallel. Today, many systems provide options in menus, direct manipulation, and command language. The resulting variety complies with the different learning phases and preferences of users. However, such variety cannot be the sole guiding principle. First, variety requires a (perhaps considerably) higher complexity of systems (system surfaces) to be shown in parallel and to be managed consistently. Second, it requires a previous definition of the number and incarnation of the different variants with respect to interaction and presentation facilities. The developer determines them, and the user can (only) select. However, many adaptation requirements occur only during use. They result from the special features of the respective tasks or the preferences and habits of the respective persons. The users should therefore be provided with their own design facilities.

Currently, users are not prepared sufficiently for designing their systems. As far as we know, they may even use the currently available adaptation facilities insufficiently. This is due to the complicated and hardly transparent adaptation tools, the low experience of the users, their orientation to task rather than tool,[4] and the reservations of organizers who do not want to con-

---

[4] The user does not process the specific task when adapting, but leaves it temporarily to design the tools. This is a meta-task that should not take too much time and that should not be too complicated.

tribute to the rank growth of systems by individual adaptation of facilities for some specific applications.

In this situation, SAGA (German acronym for software-ergonomic analysis and design of adaptivity) tries to improve the controllability of increasingly complex applications by means of individualization features that support users by system features rather than leaving adaptation to the users alone (Oppermann, 1994a, 1994b). The aim was to combine adaptable and adaptive features. The idea of adaptive characteristics of systems originates from computer science (above all, artificial intelligence) and aims at systems that automatically adapt to the specific task and the specific user based on observing the respective use. According to the critics of this concept, such an adaptation made by the technical system itself is a threat to data and privacy protection by recording the specific user behavior, and it deprives users of the control of their applications. The SAGA project attempted to combine the advantages of both concepts such that their respective disadvantages were avoided. The objective was to estimate the efficiency of and the limits to adaptivity and to give design recommendations for practical development.

Goals and Problems of Adaptive Systems

In order to find ways to make human–machine processes more productive, current research investigates the feasibility of a new kind of application designed to act as an assistant to the user, rather than as a rigid tool. An assistant shows different behavior than do current rigid systems. An ideal human assistant adopts the perspective of the client: The assistant learns about the client and knows about the client's needs, preferences, and intentions. In human–human communication, both participants can adapt their behavior according to the characteristics of the partner. There are many initial clues about the character of the partner, the intentions to open a communication, and the style of communication. The communication process enriches and refines the knowledge of both partners about each other: They learn from the verbal and nonverbal behavior. This knowledge enables the partners to improve the communication process to accelerate the discovery of common or conflicting aims, to optimize communication, and thereby to increase their satisfaction with the interaction. So much for human–human communication—what about human–computer communication? What can a technical assistant adopt from a human assistant? Why not transfer the effective mechanism of human–human communication to the interaction between a user and a system? Why not let the system learn about the user and model the characteristics of the user in its knowledge base, in order to adapt itself to the user? There has been some research in the last 10 years on developing methods for building user models and adaptive features. These attempts did not, however, result in any final conclusions—they show, at best, anecdotal impressions. In this subsection, we discuss problems and possibilities of adaptive systems and present examples of unpretentious but successful adaptation features.

Adaptivity in the form of an adaptive systems is based on the assumption that the system is able to adapt itself to the intentions and tasks of the user by an evaluation of user behavior, thus breaking down a communication barrier between human and machine (see Hayes et al., 1981). Consequently, adaptivity is basically implemented by referring actions to action patterns and finally to action plans on which the actions of the person are based. These action patterns must be identified in the evaluation of user action steps. The interpretation of user actions serves to identify user intentions. Deducing user intentions from user actions may be done in several ways. For example, having deduced an action plan, the system may initiate a dialogue to identify the user's real intentions. The system presents alternative interpretations, and the user selects the appropriate one. In this case, the current action is checked for specific regularities agreeing with previous action sequences. Where agreements are detected, appropriate subsequent actions will be assumed. In this way, computer systems should become able to adapt to anticipated user behavior in a way analogous to human–human communication.

The goal of adaptive systems is to increase the suitability of the system for specific tasks, facilitate handling the system for specific users, and so enhance user productivity, optimize workloads, and increase user satisfaction. The ambition of adaptivity is not only that "everyone should be computer literate," but also that "computers should be user literate" (Browne et al., 1990). There are examples of adaptive systems that support the user in the learning and training phase by introducing the user into the system operation. Others draw the user's attention to tools he or she is not familiar with, to perform specific operations for routine tasks more rapidly. Evaluation of system use is designed to reduce system complexity for the user. In the event of errors or disorientation on the part of the user, or where the user requires help, the adaptive system provides task-related and user-related explanations. Correction is to be employed automatically when user errors can be uniquely identified. The user is spared the necessity of correcting obvious errors.

An adaptive system is based on knowledge about the system, its interface, the task domain, and the user (see Norcio & Stanley, 1989). It must be able to match particular system responses with particular usage profiles. Edmonds (1987) described five areas the system can take into account: user errors, user characteristics, user performance, user goals, and the information environment. He suggested that user errors are the most prominent candidate for automatic adaptations. Benyon and Murray (1988) proposed adaptations of the system to the environment of the user at the interface level rather than at the user's conceptual task level. Salvendy (1991) referred to personality traits, such as field dependence versus field independence (see also Fowler et al., 1987), or introversion versus extroversion.

There are attempts to find psychological theories to describe human characteristics that are more or less persistent and are of more or less relevance for the adaptation process (see Benyon & Murray, 1988; Benyon et al., 1990; Norcio & Stanley, 1989; van der Veer et al., 1985). To the best of our knowledge, no working applications have been developed for which behavior depends on the relationship between a personal trait and system behavior. The most solid candidate for a relationship between personal behavior (although without an underlying general personality trait) and system behavior is a system reported in (Browne et al., 1990). The adaptive feature of the system is concerned with the correction of misspellings. The users can be described by their error frequency (missing letter, extra letter, wrong letter, transposed adjacent letters) and can be supported by appropriately arranged lists of suggestions for correction. The limitation of this example is obvious. It cannot be generalized, and the advantage is limited to a reduction of some milliseconds in the correction time. The same kind of limited advantage (but without the disadvantage of dynamic menus as in the next example) is provided by adaptive menu defaults, where the mouse cursor is prepositioned at the user's most likely selection (see Browne et al., 1990). In another example, the adaptive effect is a reordering of menus, where the most frequently used menu options are presented at the top. This leads to a decreased access time but also—especially in the course of learning the system—to an increased search time (see Mitchell & Shneiderman, 1989). Proofs for the feasibility and accountability of adaptive systems with respect to personality traits, such as introversion/extroversion, or field dependence/field independence (see Salvendy, 1991), where relevant system behavior corresponds to personality differences, have yet to be shown. Considering the difficulties in developing adaptive systems that respond to specific personality traits, we concentrate on task-specific regularities that can be related to differentiated system behavior.

There is an ongoing discussion about the possible dangers of adaptivity, in the form of drawbacks and problems that are in opposition to the objectives of adaptive systems. The user is observed by the system, actions are recorded, and information about the user and task is inferred. This gives rise to data and privacy protection problems (i.e., the individual user is exposed to social control). This social monitoring possibility is aggravated by monitoring of the user by the technical system. The user is, or feels, at the mercy of the control and dynamics of the system. A system hidden from the user, takes the lead, makes proposals, and, in some cases, gives help and advice without being asked. This makes the technical system appear presumptuous and anthropomorphized. Another problem with self-adapting systems is that the user is exposed to the pressure of adapting to the developer's conceptual model, on which the system is based. Finally, the user can be distracted from the task by following modifications of the system or modification suggestions. The user's overall grasp of the system structure and system capabilities is lost and transferred to the system. The user's mental model of the system, which has been acquired in using the system, becomes confused.

In order to overcome at least some of the specified problems of adaptive systems, while still achieving their objectives, the majority of authors propose that the user control how the system adapts to user behavior. This can be done by (a) providing means for the user to activate and deactivate adaptation for the overall system or individual parts of the system before any adaptation is made or after a specific adaptation state has been reached; (b) offering the adaptation to the user in the form of a proposal to accept or reject, or enabling the user to select among various possibilities of adaptation modification; (c) enabling the user to define, in an adaptation-resistant manner, specific parameters required for adaptation by the system; (d) giving the user information on the effects of the adaptation modification, which may protect the user from surprises; and (e) giving the user sole control over the use of his or her behavior records and their evaluation.

Design and Analysis of Adaptive and Adaptable Systems

The question about the adequate design of adaptive and adaptable concepts cannot be answered globally and without an empirical basis. Adaptive and adaptable features rather have to be identified on the basis of task performing in actual systems. For this purpose, adaptive and adaptable features have been developed as examples to evaluate them with respect to their relevance to the user. The evaluation considered the demand for adaptation in a task context and the learning process of the adaptivity features proceeding in the user's mental models. The design and evaluation of prototypes were accompanied by interaction analyses of the support to be given to the user when processing action plans by means of context-sensitive help. These features were first explored in human tutorial interactions; subsequently, they were gradually integrated into human–machine interaction and finally, they were transferred into knowledge-based technical components, as shown in Fig. 2.6.

The adaptivity features were converted prototypically for a commercial spreadsheet analysis to base the relevant interface features on an advanced and complex application and to be able to process authentic test tasks.

The user interactions with the application were recorded continuously. On this basis, two groups of features were generated: hypermedia context-sensitive help and adaptation suggestions.

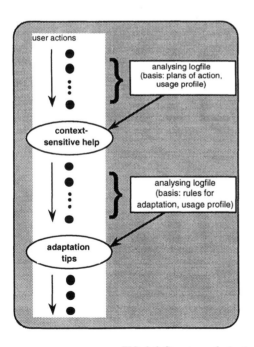

- Context-sensitive help facilities enabling users to process their problems in difficult situations were provided based on user action plans (HyPLAN).

- Adaptation suggestions enabling users a task- and user-specific design of their system were made on the basis of adaptation rules (FLEXCEL).

FIG. 2.6 Structure of adaptive user support.

HyPLAN: Hypermedia Help Based on Plan Recognition

HyPLAN is a context-sensitive help environment for spreadsheet calculations (Grunst, 1993). The system consists of two modules: the plan recognition program PLANET and the interactive hypermedia help environment HyTASK. Plan recognition techniques were used to identify possible support requirements during work and to provide selective help access. The support is mainly provided in form of animated example demonstrations.

Basic concepts for the development of HyPLAN were obtained from analyses of tutorial interactions (O'Malley et al., 1985). Experts showed casual EXCEL™ users how to solve their problems efficiently (Phillips et al., 1988). The human behavior was taken as a kind of success model for the design of support concepts. The experiments were based on the methodology of qualitative content-analyses using videos (Grunst, 1988; Suchman & Trigg, 1991). An experimental series typically began with preparatory tutorials tailored to the specific demands of the test persons. Newly acquired system knowledge was then examined by processing authentic tasks derived from the test persons' field of work.

The activities and utterances of the interaction partners as well as the computer screen were recorded synchronously. Problematic sequences were transcribed. The screen display, verbal responses and nonverbal responses, as well as activities of the test person and the tutor in their precise temporal relationship, were fixed. Thus, it was possible to identify the origin of specific (mis)conceptions. Furthermore, the structure and success criteria of complex interaction patterns such as "advice," "correction," or "critique" (Rehbein, 1980) could be reconstructed.

In order to transfer identified human success models into technical support concepts, two requirements turned out to be crucial:

- The help should mainly be realized in the form of intuitive demonstrations.
- The access of pertinent help items had to be supported in a context-sensitive manner.

The operationalization and technical implementation of the first requirement led to the development of a network of animated tutorials and help items. The support items were made accessable through an entry box, which gets adapted by the knowledge-based plan recognition component PLANET (Quast, 1993).

The features of this module were specified in previous simulation experiments applying a Wizard-of-Oz experimental setup (Hill & Miller, 1988) with human experts in the background. In contrast to the tutorial interactions illustrated earlier, expert(s) and lay person(s) were now located in different rooms. Thus, verbal and nonverbal communication were blocked. Like conceivable plan recognition systems, the experts could only derive hypotheses from inputs that could be identified at the computer screen. The utterances of the cooperating test persons and those of the experts were recorded synchronously and analyzed. Difficulties of lay persons and hypotheses of experts could be reconstructed from this twin dialogue constellation.

The reconstruction of clues and recognition patterns that were used by the experts for their plan and/or problem hypotheses could roughly be assigned to two different cognitive processes. On the one hand, a content-related understanding of the entire task situation was gradually refined, and the actual input was evaluated against this background. On the other hand, certain key operations or error messages of the system immediately triggered complex hypotheses. The first type of situational interpretation was often based on information such as the content evaluation of column headings. These inferences are difficult to simulate with a technical system. However, the second type of information—identified cue actions—can be exploited for the immediate activation of high-level hypotheses.

The PLANET Plan Recognition Component

The plan recognition module PLANET[5] evaluates recorded EXCEL operations in order to provide context sensitive help access. PLANET receives lists of macro records that document current EXCEL operations at the level of basic semantic actions. The exchange is triggered in fixed time intervals and by help calls of the user. PLANET itself puts analyzed action goals into a structured browser that serves as an entry box to HyTASK help items. Figure 2.7 comprises an overview representation of the PLANET architecture within HyPLAN.

The EXCEL macro record is preprocessed by a simple "parser" (structure interpreter), which adjusts the input strings for the inference component. The major process continuously updates seven blackboard layers of identified

---

[5] PLANET was developed in CLOS (Common LISP Object System) on a MacIvory, an Apple Macintosh featuring a Symbolics LISP board. There are two links to Macintosh applications, EXCEL and HyperCard (constituting the HyTASK help environment).

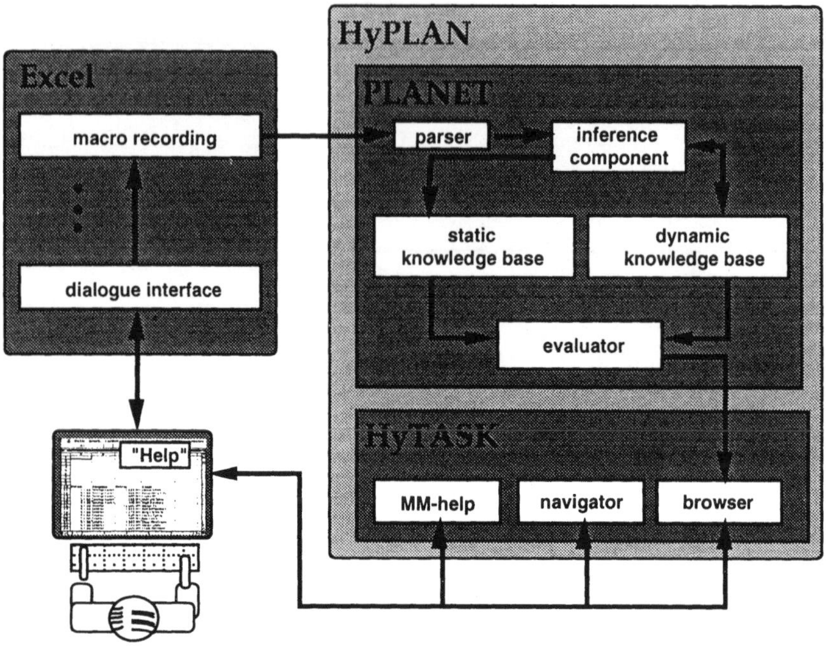

FIG. 2.7 PLANET/HyTASK within the HyPLAN help environment.

operations on different complexity levels. A spreading process creates object instances of action types that contain recognized operations as elements. This is done according to the hierarchically arranged action plan modules of the static knowledge base. In a further process, operations defined as key operations activate complex action units that are written into separate hypothesis blackboards.

The static knowledge base contains structural information about the constituent elements of complex action units. Time-logical and content-logical constraints are assigned to object classes. The dynamic knowledge base generates and manages a history tree that coordinates activated action alternatives with continuously incoming record data. Identified operation units, that is, operation units fully confirmed in one realization variant, are written into the hierarchical blackboards. These units activate all possible action types representing superordinate action alternatives. References to the triggering unit and a time mark are instantiated in order to guide further inferences.

The Interaction Elements of HyTASK

HyTASK provides task-specific help items and tutorials in a hypermedia system (Nielsen, 1990). The items include short textual explanations as well as animated demonstrations commented by spoken text. Help can be accessed directly from a general entrance menu (Navigator) or from the Browser selection adapted by PLANET. Figure 2.8 gives an overview of the different interaction components in HyTASK.

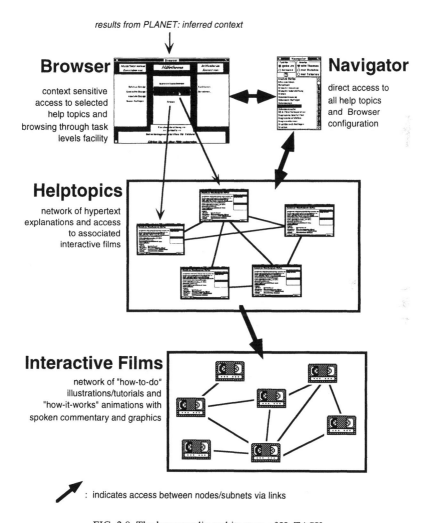

FIG. 2.8 The hypermedia architecture of HyTASK.

The help items are presented as dialogue boxes, text cards, and verbally commented animation sequences (Palmiter & Elkerton, 1991) that demonstrate pertinent actions. The help topics impart both "how-to-do" and "how-it-works" knowledge (Kieras & Polson, 1985). In particular, tutorials contain different explanation levels and control options. There are click-sensitive icons and text fields allowing switching between (further) help topics and tutorials. Related illustrations, animations, and/or texts at different explanatory levels can be accessed from relevant help contexts.

FLEXCEL: "Flexible EXCEL" for System-Initiated and User-Initiated Adaptation

FLEXCEL (Thomas, 1993) is an extension of EXCEL to an adaptive and customizable system. The primary aim of this extension is to provide a convenient environment for adapting EXCEL's user interface to particular users and their current tasks. FLEXCEL allows the user to define new menu entries for actions that are frequently needed and normally require a tedious dialog step (adaptable component). Moreover, FLEXCEL analyzes the user's interaction style and presents adaptation suggestions (adaptive component). A critique component supports users in their problem-solving and learning activities related to the adaptability of FLEXCEL's user interface.

FLEXCEL consists of two components: (a) an EXCEL component with the user interface, and (b) a knowledge-based component responsible for generating adaptation suggestions (called "tips") and critique. The EXCEL component has been implemented in the EXCEL macro language, the knowledge-based component in CLOS. All user actions are recorded in a logfile and then sent to the knowledge base. Then the rules in the knowledge base are activated; this may lead to an adaptation tip or a usage tip which is then transferred back to EXCEL. Like HyTASK, FLEXCEL was developed on the MacIvory. The EXCEL component of FLEXCEL with its user interface is running on the Macintosh, and the knowledge-based component is running on the LISP Machine. The two components communicate via file transfer. Figure 2.9 gives a simplified representation of the FLEXCEL structure and data flow.

The user interface is almost identical with that of the (nonadaptable) EXCEL. The normal functionality is extended by adaptability supporting the definition of key shortcuts and menu entries for function parameterization. The advantages of being able to adapt EXCEL in this way are obvious. The user of a spreadsheet program like EXCEL is frequently faced with tasks that

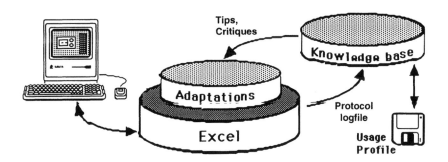

FIG. 2.9 Structure and data flow of FLEXCEL.

require execution of a function repeatedly with the same parameter(s). An (EXCEL) example may illustrate this: Let's say a user frequently needs to delete the formats in a set of cells. This normally requires selecting the function "Delete contents" from the menu, then specifying the parameter "formats" and clicking an "OK" button in the dialog box. By using FLEXCEL, the user can eliminate this dialog step by defining a new menu entry and/or a key shortcut for "Delete contents/formats." To put the adaptations under user control, the usage profile and the adaptations made with FLEXCEL can be stored on an external floppy disk. This profile concept separates the resulting adapted system from the application and from the user documents. Thus users have the control over their personal profile, which may have sensitive information about the working style, preferences, fuzziness, errors, and so on.

For the adaptive component of FLEXCEL, the user actions are kept in a logfile and interpreted by the knowledge-based component of FLEXCEL. The usage profile contains all data appropriate for generating adaptation suggestions. This includes a description of the current state of the FLEXCEL system (e.g., all user-defined menu entries), data about the usage of different functions, and data about the user's interaction with the adaptation environment. Rules attempt to identify repeated function calls with identical parameters. The continuously updated usage profile ensures that the recommendations in the tips are tailored to the specific user: If, for example, the person is identified as a typical menu user (he or she mainly uses the menu for function references and no shortcuts), the adaptation tip recommends defining a new abbreviation for the current parameterization.

Other rules can generate usage tips. A usage tip draws the user's attention to an already defined setting (e.g., a menu entry) that he or she should use for executing a function instead of going through the complete dialog sequence.

The critique rules are triggered only when the user asks for critique. The given critique is restricted to the user's interaction with the adaptation tools. The aim of the critique module is to give the user hints as to how the tools might be used more efficiently. For example, if the user continuously defines new shortcuts but does not use them or if the user makes many mistakes when using shortcuts, the user receives a recommendation to define menu entries instead of (or in addition to) shortcuts. If a user seems to have no problems with the tools, there will be no criticism presented.

The user has access to the adaptation tools in the dialog box of each adaptable function. By clicking the button "With adaptations," the dialog box expands to a combined execution and adaptation box. (Figure 2.10 shows the extended dialog box for the function "Paste special" as an example.) The adaptation part of the dialog box is basically the same for all adaptable func-

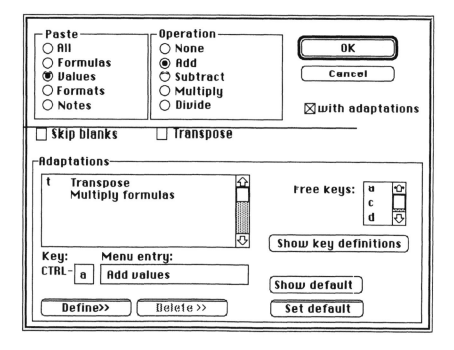

FIG. 2.10 Combined adaptation/execution dialog box for function "Paste special."

tions. The user may enter parameters in the upper part of the box, and then may define a key shortcut and/or a menu entry for the specified parameterization. The user can make definitions for several different parameter combinations without leaving the dialog box (this is symbolized by ">>" on the "Define" button). The user can select an available key from a key list or directly type in a key. He or she can inspect a list with the actually defined entries for parameterizations within the function at hand and can call up an overview of shortcut definitions of the entire system. In addition to this local access to the adaptation facilities, the user is also provided with a global access via the "adaptation tool bar". This adaptation tool bar is designed as a generic tool enabling the user to adapt the system. It contains buttons that should combine adaptive and adaptable features. Figure 2.11 describes the features in brief.

- Adaptive suggestions are presented as "tips" and are indicated by a tone, three blinks of the button "The Tip" in the adaptation tool bar, and a subsequent corona around the icon. In order to read the tip, the user can click the button.

- Unread tips are collected in a "Tip List" so that they can be accessed by the user at any appropriate time.

- The button "Adapt" presents to the user a list of all adaptable EXCEL functions. The functions can be selected, new adaptations can be defined, or existing adaptations can be changed by the user.

- The button "Overview" displays a summary of all user-defined keystrokes and menu entries.

- The "Critique" button may be clicked whenever the user wants his or her interaction with the adaptation environment to be critically analyzed. The user is told in what ways the adaptation tools can be used more efficiently.

- Clicking on the "Tutorial" button starts a tutorial on the adaptation facilities. The tutorial is an interactive animation explaining the basic ideas of how to adapt the interface.

FIG. 2.11 Flexcel adaptation tool bar.

Evaluation of HyPLAN and FLEXCEL

The tests of HyPLAN and FLEXCEL were carried out in multiple cycles of design, evaluation and redesign. They have shown that help and adaptation facilities are used only under specific conditions:

Task accomplishment should not be interrupted by a system intervention in the form of a help function or an adaptation suggestion. Users should always be able to decide themselves whether help or adaptation is just necessary or reasonable. They should at best obtain unobtrusive indications to support potentials which should however not force them to interrupt their action plans. In a preliminary version of FLEXCEL, an adaptation suggested by the system was presented by a message in the dialogue process and the user had to answer this message. Although the user was able to reject the suggestion simply by a mouse click, this form of active presentation was not acceptable. The necessity of an explicit response to specific suggestions at specific times interrupted the user's work flow. The approach selected in the second version of FLEXCEL, namely, a concurrent display of tips by means of a combination of acoustic and optical signals, has proved to be much better. The HyPLAN help component has always been passive.

Providing help and adaptation facilities exactly when required is not really feasible. According to our empirical experiments, even human advisors are not able to identify contents and time of support requirements exactly if their information basis is restricted to the information basically available for the computer. A support should therefore attempt to define the requirements of the user as precisely as possible in the form of a spectrum of possible alternatives, but it should also make the user familiar with the possibility of detail selection and further branching.

The user's control competence should not only be preserved with regard to the decision about time and content of the support suggestions—going beyond the possibility of a mere rejection of suggestions. The users should also activate help for topics of interest and should be able to perform adaptations without having to wait for suggestions from the system. In this sense, adaptive adaptation tips can be understood as a guide to adaptation through the user.

Adaptations through the user should be supported in general. At least at the beginning, the system should support the user in the following:

- Recognizing the rationale of tailoring facilities (reasons, contents, consequences).

- Practical performance of adaptations (provision of environment with selection/definition facilities etc.).
- Obtaining an overview of adaptations.
- Modifying adaptations.

Adapting is no one-shot action. It is only done after some period of familiarization with the relevant application and one's own tasks. It is often an iterative process; however, sometimes it is also of an excursion character, which has to be supported by a facility for consistent execution and control of adaptation.

Creating and naming of adaptations can become the subject of consultation. Users are sometimes able to criticize their own measures when confronted with their result within the system context. A human advisor is sometimes also required, both for the creation and the management of individual adaptations and even more for design of adaptations to cooperative work contexts. A critique can sometimes also be left to the system, as is done in the critique component of FLEXCEL with respect to the utilisation of adaptation facilities or in HyPLAN with respect to system functionality. In this context, moderation, adequate formulations, and presentation are even more important than in the case of suggested adaptations (constructive critique, multimedia presentation).

The adaptation results should be fitted into the available application to form an integral part. The creation of a separate set of individual commands in the form of an additional menu (cf. "Mod." menu in Microsoft Word on Macintosh) or an additional application-specific tool bar (cf. "button bar" in Word-Perfect for DOS computers) is not the correct approach because it creates a new world in addition to the familiar one and requires an unnecessary additional training effort. In FLEXCEL, the additional user-specific commands have been indented in the normal menu under the corresponding functions so that the test persons had no difficulties in finding and using them if necessary without any previous explanation.

Adaptations should be coordinated with the user's different task classes. When using a generic application, users do not necessarily work on one single task and should therefore be able to create specific working environments for different tasks using the underlying application. Such a classification of adaptations can also be helpful as a basis for the design of application configurations for working groups. This facility is provided in the FLEXCEL adaptation concept by a concentration of adaptations in a "profile" that the user may save on a floppy and load if required by selecting from a choice of available profiles. The current profile is displayed in the header line of the main menu. A minimum data protection of sensitive user data is thus achieved by recording the profile on a floppy disk and by its separation from the application—the data are therefore not available for a central evaluation of user characteristics.

The work in the SAGA project has shown that the often discussed difference between adaptability and automatic adaptation by the system is misleading. Both approaches have their weaknesses if pursued separately, and they should be implemented in combination. Approaches to do so have been developed and tested in the prototypes HyPLAN and FLEXCEL. In addition, the work has shown that a sole orientation of adaptation to the individual user in the sense of an individualization can lead to an isolation. Instead, adaptation should be possible in and for groups and should support cooperative work through interchangeability and through management of group-specific and person specific adaptation configurations (profiles). In this fashion, adaptation can make a contribution to coping with the increasing system complexity in organizations.

Acknowledgment

We are grateful to Thorsten Fox and Klaus-Jürgen Quast for their valuable contribution to the subsection about HyPLAN.

2.3 PROCESSING INEXACT KNOWLEDGE

Franco di Primio

In this section we concentrate on one assistance property: the ability to handle inexact knowledge and instructions. Research on this point was done at GMD within the TASSO project[6] (TASSO is an inexact acronym for Technical Assisting Systems for Processing Inexact Knowledge). The main goal was to study, contrast, and develop new methods for handling incomplete, vague, or contradictory knowledge. From an interactive point of view, which is relevant for the quality of the assisting modalities, it is important that application systems incorporating these methods must be decidedly more flexible than conventional ones, with the ability, for instance, to interpret inexact user instructions. As an example application field, we have studied the problems of processing imprecision during the construction and search of graphics. The user is expected to incrementally specify the attributes of graphic objects (such as presentation graphics or diagrams of office furniture layouts). Adequate assistance consists here in completing and making more precise or consistent user specifications that are inexact in the preceding sense. The techniques we have studied and applied are mainly based on nonclassical inference, such as nonmonotonic and associative reasoning, and on methods of planning under uncertainty.

The study of nonmonotonic inference mechanisms (Brewka, 1991) is of central interest because they allow the assisting system to handle incomplete problem specifications on the basis of standard assumptions. Default rules expressing typical but not universally valid relationships help, for instance, to infer plausible conclusions and take meaningful decisions in the absence of precise knowledge.

Associative reasoning techniques based on neural networks (Henne, 1991; Paaß, 1992) realize a strict memory-based approach and are being used in this sense for the storage and retrieval of known problem specifications and solutions. To solve a problem in this context means to look at similar, already solved problems. These are reactivated in a content- and context-driven manner.

---

[6] The project TASSO was partially funded by the German Federal Ministry for Research and Technology under contract no ITW8900A7.

Research on planning under uncertainty is motivated by its special role in the construction of graphics. The design of semistandardized graphics like business presentations can be viewed as a configuration task. In this context, existing techniques for planning and configuration must be extended to allow the inexact specification of the planning model, the planning operators, and the goals (Hertzberg, 1991).

Finally, since a problem specification may be inexact in every respect, that is, incomplete and/or vague and/or contradictory, the problem arises of how to apply the different methods and inference techniques in a coordinated and complementary manner. This we expected to be a problem of designing an effective and efficient hybrid architecture, that is, one allowing the use, at the right moment, of the right method for the right (sub)problem (di Primio, 1993).

In the following subsections we outline the main results we have so far achieved in the study of nonmonotonic and associative reasoning techniques. In this presentation, we take a purely functional perspective, meaning that we want to stress the aspects of the methods that are relevant from the point of view of an end user, omitting the (considerably more but partly tedious) details concerning the realization of the techniques (for which the reader is referred to the technical project reports). Furthermore, we completely skip the research on planning under uncertainty. The section by Hertzberg in this book is concerned with this problem.

Why nonclassical inference methods

The basic motivation for studying and applying nonclassical inference methods lies in the strong need for flexibility in order to handle inexact knowledge. This is best understood by reference to the notion of a frame, which, as a major type of knowledge representation structure, has been, toward this goal, extremely influential in artificial intelligence (Minsky, 1980). The premise underlying the frame idea is that knowledge about regularities in the world (such as the likely properties of objects and situations) is stored in the memory in clusters that can be accessed at need as large, coherent units and that can serve to generate plausible inferences ("expectations") helping to fill missing details in a given situation. Frames are paradigmatic for the idea of memory-based reasoning, that is, reasoning that is not the result of sophisticated logical inferences but rather of recalling and exploiting known situations. Perhaps their value is best summarized by a (provocative and paradoxical) sentence of C. F. von Weizsäcker, "Man sieht nur, was man weiß" (We see only what we know)[7], which, in our context, can be used to emphasize the intended expectation-driven function of frames in cognitive processing. This function is crucial as it allows completion by "default" values (that are assumed unless explicit information is supplied to the contrary) for partially specified or accessible situations.

As devices for chunking information, frames have, indeed, originally inspired a bunch of productive research and application in default reasoning. Although they have been demonstrated to be very valuable in formalizing knowledge about recurrent and stereotypical situations that can be used for "completion" purposes, they show some deficiencies in other respects. They are, for instance, not very flexible and sound in handling exceptions. What is to be done, in the course of a problem-solving process, with objects or situations that are not properly matched by any known (stored) frame? Another problem can be called the "deactivation" (or "revision") problem. Suppose a frame has been activated (accessed) in response to partial aspects, that is, some features of a given situation, and suppose that it turns out to be the "wrong" frame (that is, the expectation is not met). How can it be deactivated? This is a complicated problem when you consider that frame activation technically can "trigger" the execution of several (so-called attached) procedures that may cause, as a side effect, the attributes of the actual and other frames to be changed in form of a "chain reaction."

The EXCEPT System

Logically, the activation (or instantiation) of a frame corresponds to making assumptions or asserting some sentences. In this view, the deactivation problem just mentioned consists in undoing the effects of activation, that is, retracting the sentences and all their consequences. The reasoning system EXCEPT (Junker, 1991) was conceived and developed with the explicit goal of correctly handling the completion problem, as well as the exception and retraction (deactivation) problem, in a logic-oriented framework.

---

[7] We quote from Bischof (1989, p. 79). It is worth noting in passing that the German word *wissen* (to know) is related to the latin verb *videre* (to see). The sentence then reads, We see only what we saw, that is, in this respect, to see can be considered to mean to remember (think of Plato's teaching) what one has already seen. Currently, the notion of "perception" is still problematic. Wittgenstein's distinction between *seeing* something and seeing it *as* something, as well as the *theory-ladenness* objection of Thomas Kuhn (see Bechtel & Abrahamsen 1991, p. 159f) strengthens, for instance, the analysis that what we see depends on what we know, and considerably challenges the notion of objective observation, jeopardizing, in this respect, the epistemological status of science.

The system provides a basic representation formalism that is almost as powerful as a first-order language. Compared to the well-known Prolog programming language (Clocksin & Mellish, 1984), not only definite but full Horn clauses (true negation) are supported. It is even possible to define general clauses (containing more than one positive literal), with the restriction, however, that variables appearing in positive literals of a clause must also be used in at least some negative literal of the same clause. In addition to this language, which makes it possible to write formulas having the character of premises, the user is also allowed to qualify formulas, that is, to characterize them as having the status of *assumptions* or *defaults*. The difference between assumptions and defaults is as follows:

Assumptions are possible hypotheses that are introduced in order to explain observations. Given, for instance, the premises $a \rightarrow b$ and (the observation) b you can hypothesize (abduce) a as an explanation of b. Abduction is clearly unsound from the deductive point of view but is a useful form of common sense reasoning. For an accurate analysis of the relationship between abduction and deduction the reader is referred to Console et al., (1991). Defaults are conjectural rules (of thumb) that are applied whenever there is no evidence against their application. An (abstract) example is: *Generally, objects of type T have property P. If A is an object of type T, then, in lack of any counterevidence, you can 'deduce' (expect) that A has property P.*

Compared to abductive assumptions, the basic function of default rules, which has probably also to do with some form of "cognitive economy", is different. Although people normally trie to minimize the number of hypotheses (assumptions) needed for explaining something, they try, in their actions or "expectations" about the surrounding situations, to maximize the number of applied defaults (this reduces the amount of missing information).

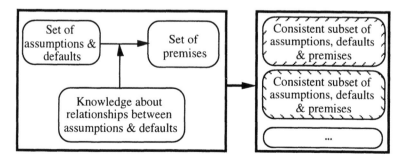

FIG. 2.12 Controlling the extension of knowledge.

Assumptions as well as defaults (like "prejudices" in everyday life) are used frequently in the reasoning process but are (hopefully) withdrawn if further premises become known that contradict some of them or some of the consequences based on them. The internal machinery used to this end by the EXCEPT system is a rule interpreter extended by reason maintenance capabilities. This combination (reason maintenance plus a system for applying rules with additional tests) is a distinguishing feature of EXCEPT and has proved powerful enough to run on the system not only various forms of default logic but also other types of nonmonotonic logics like, for instance, autoepistemic logic (Junker 1991, p. 37f)[8].

From the user's point of view, the interesting aspect is that assumptions and defaults, of course, are not unrelated. They may depend on each other. They can, for instance, be more specific than others or conflict with others, where exceptions are a particular form of conflict.

To understand the general perfomance of the system, it is useful to consider the set of premises together with the set of formulas expressing assumptions and defaults. To the same degree as assumptions and defaults are in conflict with each other and the premises, they make the resulting set (union) inconsistent. From this abstract point of view, the output of the system consists in partitioning this set into subsets that are maximal with respect to consistency and that satisfy the additional restriction to contain as a subset the set of premises, that is, as much of the available exact information as possible. Each of these maximal (largest) consistent subsets (MCS for short) represents then a possible completion of the premises. This is graphically summarized as in Fig. 2.12.

Here the knowledge about the relationships between assumptions and defaults (including priorities or partial orders [preferences]) is used to control the extension of the set of premises through instances of defaults and assumptions. This (second-order) knowledge can be expressed as a set of meta-predicates (a meta-language) ranging over formulas of the basic representation language (the object language). The performance of the EXCEPT system can then be viewed and explained logically in terms of the properties of the resulting amalgamated language.

---

8 Autoepistemic reasoning is a further form of nonmonotonic reasoning that is revisable because it is of introspective nature (*From my current state of knowledge, I can deduce that* ...), that is, it is "valid" only with respect to the state of knowledge of a particular reasoner, which is subject to change (Thaise, 1988, p.165ff).

From an incremental and interactive point of view one can see the process of constructing MCS as maintaining multiple contexts, where each context represents an alternative completion of a partial problem specification to be offered to the user as a choice. A satisfactory management of multiple contexts, including the possibility for the user to undo the effects of previous choices, is not an easy task. It is, however, essential because, for instance, it allows the user to explore the consequences of different instructions. A further problematic point consists in determining the right step size in the successive construction of an MCS, which means to be able to anticipate the desires of the user not as much as possible but as much as needed by the actual context, that is, completing gradually and not all at once what is left unspecified. A too far-reaching anticipation could in fact cause a lot of retractions to be done if the user intended to follow a different construction path (see di Primio, 1991, p. 21f, for further considerations).

Associative, Memory-Based Reasoning

The EXCEPT system offers basic representation primitives (premises, assumptions, and defaults) that can be incrementally constructed and retracted so as to serve as flexible building blocks for the specification of complex domain and user preference models. If we extend, nevertheless, to additional representation and inference techniques, we do so for many different reasons. A first general reason concerns the problem of handling global versus local inconsistency. In the logic-oriented framework, inconsistency is a local phenomenon with global effects. In other words, having, for instance, two contradictory clauses in a set of clauses renders the whole set useless in the sense that everything can be deduced from it (this is the classical Aristotelian principle *ex contradictione sequitur quodlibet*). Such a situation can, in general, only be "repaired" by isolating all the contradictory subsets, which is a potentially exponential time-consuming task as one has to consider, in the limit, the whole power set of a given set of clauses[9]. Handling inconsistency in a frame-based representation is, in general, less costly. Frame structures do not stand for isolated but for highly interrelated chunks of knowledge, connected, for instance, through specialization (inheritance) links. Doing inferences consists in traversing paths along such links. As frames can be focused on individually, they provide flexible entry points into

---

[9] Note that computing *minimal (smallest) inconsistent* subsets is the inverse problem to the problem, mentioned earlier, of computing the *maximal consistent* subsets of a given set of clauses.

a net. This means that one seldom has to consider the whole frame network and, depending on the form of a subnet and the type of links involved (Touretzky et al., 1988), one can be sure to make, within it, correct inferences without having to care for other, possibly inconsistency bearing parts of the same network. This local safety is, of course, gained at the cost of a more limited inference capability. Traversing links is, on the other hand, much more efficient than doing resolution steps in a logic-oriented language. But more than efficiency, the basic point here is that the need for global consistency is reduced or, loosely speaking, the bite it is taken out of it.

Other important reasons for trying alternative, frame-oriented forms of knowledge representation are related to the properties of the application field. The point to note here is that we have to cope with objects and situations that have an "extension" (can be represented graphically) and that are related to each other in the sense of part–whole relationships. It is generally agreed that handling complex compound objects, i. e. objects composed of parts, each part being either another complex or a primitive object, is not easy for logic-oriented representation languages like EXCEPT. They have, in this respect, the same problems as those known in the context of relational database languages. If required, a complex object can here only be represented by a multitude of clauses (tuples). In this way, the object is, as Reimer (1989, p. 11) puts it, "absolutely non existent" and can only be "(re)assembled" from the different parts through appropriate user queries. That means that the responsibility for the reconstruction of a complex object lies with the user and not with the (database) system. A frame-oriented representation does not suffer from this problem. Another point is that reasoning about extensions means, in the extreme case, to be capable of doing analytical geometry, which is a very complex (time-consuming) task in a pure logic-oriented framework. Our intention, in this respect, is even to dispense completely with analytical geometry and to study and apply, instead, what we call *model-based* methods of spatial representation and reasoning.

Model-Based Versus Syntactic Reasoning

Because doing spatial or graphic reasoning in a model-based way is a basic concern for the project, we give here a short example that illustrates our understanding of the expression "model-based" and that helps to clarify the underlying motivation. Model-based reasoning is best viewed in contrast to purely propositional (syntactic) reasoning. Here is an example of propositional reasoning:

Consider two different sets (M1 and M2) of objects, where the elements of M2 are sets that contain elements of M1. Suppose the following is known to hold:

- Any two members of M1 are contained in just one member of M2.
- No member of M1 is contained in more than two members of M2.
- The members of M1 are not all contained in a single member of M2.
- Any two members of M2 contain just one member of M1.
- No member of M2 contains more than two members of M1.

From this small set of premises we can derive a number of theorems by using customary rules of inference, such as are implemented in systems like EXCEPT. It can be shown, for instance, that M1 contains just three elements. Is the syntactic transformation of premises by rules of inference the only way of making explicit what is implicit in them? Another possibility consists in building a geometrical model that gives a "concrete" spatial interpretation, that is, a semantic, for the symbols (naming variables and relations) used in the premises. Take, for instance, the elements of M2 "to be" the sides, and the elements of M1 to be the vertices of a triangle, as in Fig. 2.13

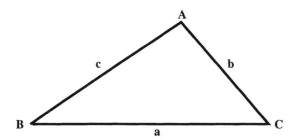

FIG. 2.13 A geometrical model for sets of objects (see text):
M1 = {A, B, C}; M2 = {a, b, c}; a = {B, C}; b = {A, C}; c = {A, B}.

Then you "see" that the preceding theorem holds[10]. There is evidence to suppose that most of the reasoning human beings do (not only in common situations) is of a similar form, that is, mental images (or, in general, models) play a decisive role (Narayanan & Chandrasekaran, 1991). They are, possi-

---

[10] We borrowed and adapted this example from Nagel & Newman (1959, p. 16ff), where it was used with the different intention to show how models can help to establish the consistency of a set of premises.

bly, the original (in the [chrono]logical sense) way to tackle problems. As another, more everyday example, the reader is asked to think over what he or she will do when requested to give an answer to the following question: How many windows are there in the front side of the house you live in? Probably, the reader will build a (mental) image of the house front and then count the windows.

Mentals models have been carefully studied by Johnson-Laird (1983) for finding answers to questions like "What happens when we understand a sentence?" or "How is it possible for you to make a valid deduction even if you have not learned logic?". The same author summarizes (1988) the difference between propositional reasoning, that is, working with formal rules that are purely syntactic, and model-based reasoning (working semantically on the basis of "vivid images") as follows:

> One way in which a valid inference can be made is to imagine the situation described by the premises, then to formulate an informative conclusion which is true in that situation, and finally to consider if there is any way in which the conclusion could be false. ... To imagine a situation is to construct a "mental model" ... based on the meaning of the premises, not their syntactic form, and on any general knowledge triggered by their interpretation. (p. 226f)

Of course, such a model-based way of proceeding perfectly fits the memory-based approach. Our basic research goal is to devise computational mechanisms that "mimic" such mental processes. In this respect, we believe that (high-level) frame-based structures supplemented by (low-level) depictions, as a mean of realizing spatial models (Habel, 1990), are a good starting point insofar as they allow one to handle knowledge of various levels and to choose different degrees of granularity for the representation.

The ASM System

A first concrete result toward the goals described above is the Associative Memory Model (ASM) of Henne (1991), a flexible experimental system that tackles different problems at different levels. Its overall organization is as follows (Fig. 2.14). The basic level consists of a neural network package that realizes the minimal functionalities needed to build general neural networks. The focus lies here on the requirement for minimality and on taking care not to be completely unrealistic from a biological point of view. The second level offers various forms of associative memory structures like object–attribute–

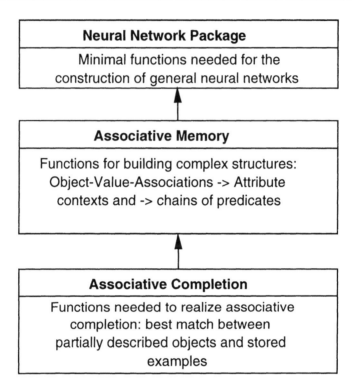

FIG. 2.14 Structure of ASM.

value triples or chains of predicates. Functionally, the user has the possibility, for instance, of inputting triples and retrieving them using attributes as context-selecting devices. The important thing is that these structures are all based on the same primitive structures (nodes) and operated on by the same basic "inference" method, that is a relaxing, value-passing, spreading-activation mechanism. Answers in the system are modeled as a network of mutually reinforcing nodes. At the third level, these structures can be used for retrieval and associative completion. Previously stored (memorized) examples of objects and situations are retrieved on the basis of partial descriptions. Means for computing best matches (such as the intersection of attribute values) are also realized using the same spreading-activation mechanism.

A basic and still unresolved issue we are working on is the handling of incrementality with respect to the following questions: How do successive inputs merge in the memory and build abstract structures (i. e. the [old and well-known] problem of learning object classes from examples)? What does "abstraction" at different levels mean, and especially, what does it actually mean to have abstraction in an extensional representation form? Supposing there is an answer to this problem, how can exceptions be handled? Furthermore, what are, in this framework, the meanings of negation, disjunctive information and inconsistency? As you can see, these are, besides the learning problems, almost the same questions that have been asked and answered in the context of nonmonotoning reasoning à la EXCEPT. The starting point is, of course, very different (only minimal conditions are assumed beforehand). The intention is different too, because the goal here is to remain as much as possible at the level of an explicit and extensional representation of knowledge.

In this respect, our view of associative, model-based reasoning joins aspects of research currently done in the context of so-called *vivid reasoning* (Etherington et al., 1989). In the broader context of hybrid reasoning this approach contrasts with other attempts looking at limited inference or syntactic restrictions on the representation. The working assumption underlying this research is that some kinds of (fast) commonsense reasoning are best modeled by simple database-style lookups. Starting from facts that may be presented in a more expressive (first-order) language, a vivid knowledge base (KB) is constructed such as to contain only elements (ground, atomic facts) that are in a one-to-one structural relationship to the parts of the world being modeled. In this way, a vivid KB becomes, like a picture, an *analogon* of the domain, such that everything that is represented is explicitly represented: "The notion of vivid representation is appealing for reasons beyond supporting reasoning as database-style lookup: it corresponds well to the kind of information expressed in pictures" (Etherington et al., 1989, p. 1147). An interesting point, in this respect, is how incomplete, for instance, *disjunctive* information is "vivified". When a given information asserts that a particular individual has one of several properties, without specifying which (e. g. "Joe is teacher *or* professor"), then a property is sought that subsumes the mentioned properties (this can be done with the help of a lattice of predicates providing subsumption information) and is used to assert in the vivid KB a ground (atomic) sentence (e. g., "Joe is instructor"). This amounts to substituting *vagueness* (the term instructor is more general than (i. e., subsumes) teacher and professor) for *ambiguity* (the disjunctive information) (Etherington et al., 1989, p. 1148f). Currently, the authors do not have a satisfactory treatment for the "vivification" of *negative* information. A further idea in this context, which is also our (long-standing) idea while trying to understand the transitions from associative to other forms of reasoning and the principles that make them possible, is that vivid (associative) reasoning has to supplement but not to replace a first-order one. In this sense, if the result of querying the vivid KB (the associative memory) yields insufficient information, one can then query the first-order (EXCEPT) KB, where more powerful but less efficient reasoning methods are used.

Autoassociative Completion

A second result in associative reasoning tackles the problem of representing and handling spatial knowledge on the basis of neural network techniques (Paaß, 1992). After discussing different forms of local and distributed representation, the author shows how to combine the ideas and intentions underlying object centered representations and feature maps. While feature maps are based on a fixed global coordinate system in space and consider the contents of each element of a discrete grid, object-centered representations describe the locations of objects relatively, that is, objects are described with respect to some variable, possibly nonuniform grid centered on another (reference) object. The problem with object-centered representations is that one has to cope with a normally variable number of objects in the neighborhood of a given object. On the other hand, the apparent inflexibility of feature maps can be (partly) remedied using overlapping receptive fields that are distributed over the map(s). If the receptive fields are big enough, they will cover most of the relevant relations between objects, that is, those depending on their immediate neighborhood. This way, the associative procedures for the different fields may be identical, that is, have the same parameters regardless of their location. The main advantage here is that fields of the same size are easily translated into an associative model with fixed input and output vector. The author shows how, according to this view, invariance with respect to shift and rotation may be achieved.

A possible extension of this approach consists in using several feature maps with different granularity or resolution and/or receptive fields of different size. This amounts then to using in the background several neural networks, where the (architectural) problem arises of how to compose (connect) them. A first step toward a solution could be to proceed along the lines suggested by Feldman (1987) (see also Paaß, 1992, p. 41ff), that is, to relate the maps to reference frames (like retina and head) having a different "positional" value in the context of a more complex and "complete" cognitive system, a system that is not only able to "see" but is also able, for instance, to "move" and have "knowledge." The reference frame labeled "world knowledge network" is indeed meant by Feldman (1987) to represent the agent's general knowledge of the world, including aspects independent of space or vision. This reference frame could be realized by or act, in our context, as an interface to other, higher level reasoning subsystems, like ASM or EXCEPT.

Examples of Interaction

We would like to give the reader a concrete feeling for the meaning of the assisting capabilities achieved so far. As application example, we take the specification of office furniture layouts. A concrete task, in this context, would be to select and arrange equipment for an office such that two office workers can do their work. This can be viewed as a complex design problem. The complexity lies in the necessity to consider different types of (quantitative, qualitative, and spatial) information, as can be appreciated by the following short characterization of possible requirements:

- The office should have at least one telephone.
- The costs for the furniture should not exceed 6000 DM.
- The desks are to be placed in such a way that the workers can directly see each other and that they enjoy optimal lighting.
- Desks and filing cabinets should be surrounded by sufficient free space.
- The distances between desks and filing cabinets should be as short as possible.
- No object should block the way to a door or a window.

The requirements clearly range over very different aspects: ergonomic, financial, positioning (spatial), and even social. To the same degree that these can be given different weights or priorities, they can be used to build different preference models. In our opinion, it is of no (very intelligent) use to consider all these aspects together and at the same time so as to face a "huge," unstructured constraint problem that one could then try to solve "monolithically" with one optimization procedure. Instead, we want different subsystems (reasoners), which handle different aspects, to cooperate in solving the whole problem. Besides this point, which justifies our different methodological approaches and developments, the central problem is how to cope with the inexact specification of such requirements. (What is the meaning of "sufficient free space"? What does it mean to enjoy optimal lighting? Are there requirements that contradict each other? etc.)

We now give an example of our actual understanding and handling of these aspects in form of three snapshots of the interaction with an office layout assisting system.

Handling Contradictory Instructions

The first snapshot concerns the handling of contradictory requirements. Suppose the design of an office layout has progressed as shown in Fig. 2.15.

This is (except for the words with outgoing arrows, which are annotations for a better understanding of the graphic symbols) the current interface of our

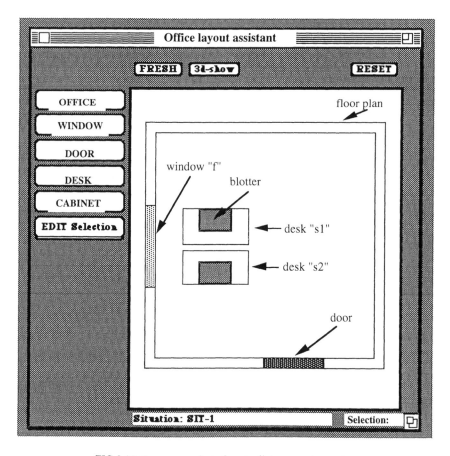

FIG 2.15 Screen snapshot of contradictory requirements.

office layout "assistant"[11]. Note the position of the blotters, which indicate the side of a desk, where a worker would sit. Desks, windows, and doors are objects that have an orientation in the sense of an intrinsic reference frame (Retz-Schmidt, 1988), that is, they have (like people) a front, back, right, and left side (in the two-dimensional representation top and bottom side are not taken into account). In this situation, the user requests the system to change the layout in such a way that some (further) requirements are fulfilled. The user input (preceded by the prompt "C <- :", where C means "construct") is written in boldface type.

> C <- : **((vis-a-vis s1 s2))**
>   **((optimal-lighting f s1))**
>   **((optimal-lighting f s2))**
>   **((for-right-hander s1))**
>   **((for-right-hander s2))**

These five requirements cannot (as the system will find out) be fulfilled at the same time. The first one means that the desks are to be positioned such that the workers can directly see each other (i. e., sit face to face). The second and third ones mean that the desks s1 and s2 should receive an optimal lighting from the window f. The last two requirements transmit to the system the information that the users of the desks are both right-handers. Note that this is a purely declarative form of instruction specification. The user only states which properties the desired result is required to exhibit but does not say (and does not know) how the desired result is to be obtained[12]. The answer given by the system (marked with "A -> :") is:

> A -> : In the current situation, one of the five requirements ((vis-a-vis s1 s2)) is
>   already fulfilled. The remaining four cannot be satisfied simultaneously.
>   There are two subsets of them expressing properties which are mutually
>   compatible:
>   1: ((for-right-hander s2)) ((optimal-lighting f s2)) ((optimal-lighting f s1))
>   2: ((for-right-hander s2)) ((for-right-hander s1)) ((optimal-lighting f s2))
>   Please, choose (1 or 2) ?

Thus, the performance of the system is twofold. First, it has recognized that one of the requirements is already satisfied, that is, it can be deduced from

---

[11] The interface is defined on the basis of the object-oriented language EPICT (Rome et al., 1990), which is necessary not only for graphic representation purposes but also as a common means for communication among the internal components of the system.

[12] This remark is also valid for the examples in the next subsections. Imperative elements do, however, appear if the user directly "intervenes" in the graphic representation, for instance, moving with the "mouse" an object from one to another position. In this case, the result of this action can (must) be translated into a set of propositions to enable the logical reasoner to handle them correctly. The translation is not a trivial task.

the premises (the current state of the situation). Second, it has found for the remaining requirements two maximal subsets (MCS) that can be used to extend the current situation without producing inconsistency. In order to be able to give such an answer, the system must, of course, have a lot of background knowledge and reasoning capabilities. It must know, for instance, that a desk that is to be used by a right-hander is not best positioned if there is only a window on its right side (for a right-hander, lighting is optimal only when coming from the left). Thus, placing two desks (for two right-handers in an office with only one window) "face to face" is not compatible with having them both in optimal lighting position. Further aspects of the knowledge required in this context are given in Junker (1991, p. 85ff) and di Primio (1991).

Associative Search of Office Layouts

Figure 2.16 illustrates the use of the ASM system for doing graphic search. The idea here is to do a search on the basis of queries that are (possibly annotated) sketches of office layouts. The system is supposed to already know several concrete examples of office layouts. Its task is to find among them one that matches as well as possible (if possible, exactly) the search query, that is, the sketch. The upper part of the figure shows on the left side a concrete office layout (labeled SIT1) consisting of two desks in front of each other and a cabinet on the right side of the window. The right side of the upper part shows the (internal) mechanism that captures the geometrical information of the layout.

This mechanism (which is not meant to be visible for an end user) is labeled RETINA in the figure and consists basically of a sensor field realizing a simple pixel feature map. The granularity of the map is chosen to be 16x16 units but could be varied, realizing higher or lower forms of resolution. In this way, the exact information of the example is retained in the associative memory (after saving) only in a coarse form (note that the position of the objects in the map is reflected along the horizontal middle axis).

The pixels (i. e., the black fields of the map) are parts of the different objects. We say that they have different "colors" depending on the "type" of the objects they belong to. Thus, there are desk pixels, door pixels, and so on. This type of information is also saved. Additional information, including aspects dealing with the general context in which the office layout has been produced, can also be stored. This means, after all, that a graphic situation can be annotated locally or globally with "propositional" information. In this

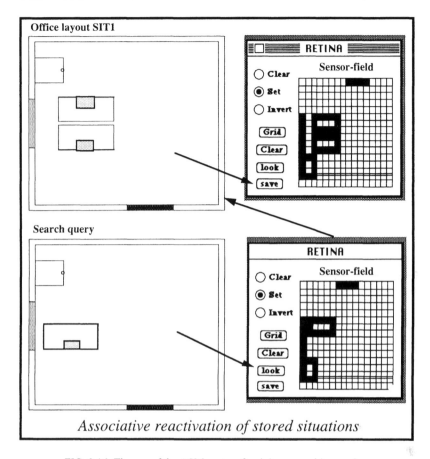

FIG. 2.16 The use of the ASM system for doing a graphic search.

respect, the ASM system has proved to be very valuable because it can flexibly deal with different representations of both aspects, the geometrical
(spatially organized) and nongeometrical features of the objects. The recall or
reactivation phase, which is outlined in the lower part of the figure, is initiated (through the look button) on the basis of a (possibly annotated) sketch of
what the user is looking for. The query is content oriented in the sense that,
besides the basic graphic symbols, no particular syntactic restrictions are
imposed for its specification.

Please note that the sketch is inexact in several respects. It is incomplete
(only some objects are specified) and imprecise in the positioning of the
objects included in the specifications. Part of this imprecision is handled

(filtered out) at the level of the RETINA because of the coarse translation in the sensor field (this can be seen as a sort of "intelligent" preprocessor). The recall strategy is actually implemented in such a way that geometrical aspects, that is, the information about the form, are considered first. If the result is insufficient (too many "hits" or no hit at all), then additional information (type and further annotations) are taken into account. In the example shown in Fig. 2.16 the search query yields exactly the office layout SIT1. We had here no big problem with multiple hits because only about ten office layouts were stored in the associative memory.

Associative Construction of Office Layouts

In this last subsection, we illustrate the use of autoassociative techniques for the construction of office layouts. To this purpose, a more flexible form of interfacing is needed. We have to be able to do associative completion locally and not only globally. Although for searching purposes it might suffice to partially specify a whole office layout and to get a (set of) known layout(s) as result, when constructing one wants, of course, to use previous examples but not to reproduce (copy) them in toto. One rather wants to adopt and adapt only parts of known layouts. In other words, a generalization capability is required. Technically, this means that one has to be able to look into the parts of existing office layouts and see and use similarities at a local level. This has been achieved using receptive fields (as explained in the earlier subsection about the ASM system). As before, the starting point is an adequate feature map where relevant information is encoded, for instance, as in Figs. 2.17 and 2.18 which come from Paaß (1992, p. 52).

Note that in Fig. 2.18 showing the encoding of the layout of Fig. 2.17, not only is type information represented but also information about the orientation of (the parts of) the objects. Thus, we have here not one but two (overlapping) feature maps. In the learning or training phase, which operates on the basis of specifications (examples) like this one or parts of them, receptive fields (of, say, 3x3 size) are built that cover the whole feature map(s). The fields may overlap and have different orientation. Instead of using many different fields in parallel, it is, of course, possible to have one receptive field scanning the map(s) sequentially. The important thing is that the contents of the field(s) are translated in the same associative network. The overlapping of fields results in an invariance with respect to shift. Different orientations are needed to achieve invariance with respect to rotation (the use of nine square pixels per field allows, of course, only a few useful rotations). In the

recall phase, partial layouts or parts (e. g., corners) of layouts can be used
that are different (in size and contents) from those used for training.

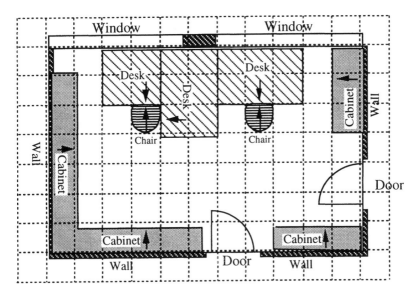

FIG. 2.17 An office with a coarse grid.

En	Fn	Fn	Fn	Fn	Fn	Wn	Fn	Fn	Fn	Fn	Fn	Ee
Ww			Ss	Ss	Sw	Sw	Ss	Ss	Ss			Rw	We
Ww	Re		Ss	Ss	Sw	Sw	Ss	Ss	Ss		Rw	We
Ww	Re			Cn	Sw	Sw		Cn				Rw	We
Ww	Re											We
Ww	Re											De
Ww	Re											De
Ww	Re	Rn	Rn	Rn	Rn	Rn			Rn	Rn	Rn	We
Ew	Ws	Ws	Ws	Ws	Ws	Ws	Ds	Ds	Ws	Ws	Ws	Es

FIG. 2.18 Active units of feature maps encoding the office of Fig. 2.17.
Terms are defined by C: chair; D: part of door; E: corner; F: part of window; R: part of
cabinet; S: part of desk; W: part of wall; blank: space; n: north; w: west; s: south; e: east.

A "construction" query like the following one (where the stars indicate missing information):

```
En Wn Fn Fn Fn Wn Fn Fn Ee
Ww  *  *  *  *  *  *  *  We
Ww  *  *  *  *  *  *  *  We
Ww  *  *  *  *  *  *  *  De
Ww  *  *  *  *  *  *  *  De
Ww  *  *  *  *  *  *  *  We
Ww  *  *  *  *  *  *  *  We
Ew Ws Ws Ds Ws Fs Fs Ws Es
```

can, for instance, yield the following completions (the minus signs indicate free places):

```
En Wn Fn Fn Fn Wn Fn Fn Ee      En Wn Fn Fn Fn Wn Fn Fn Ee
Ww Ss Ss Ss Ss Ss Ss Ss We      Ww Ss Ss Ss Ss Ss Ss Ss We
Ww Re Cn  - Cn  - Cn  - We      Ww Re  -  - Cn  - Cn  - We
Ww Re  -  -  -  -  -  - De      Ww Re  -  -  -  -  -  - De
Ww Re  -  -  -  -  - Rw We      Ww Re  -  -  -  -  -  - We
Ww  -  -  -  -  -  - Rw We      Ww  -  -  -  -  -  - Rw We
Ww Re Rn Rn Rn Sn Sn Sn We      Ww Rn Rn Rn Rn Sn Sn Sn We
Ew Ws Ws Ws Ws Fs Fs Ws Es      Ew Ws Ws Ws Ws Fs Fs Ws Es
```

A substantial problem faced here is how to allow the user to give the system instructions containing negative, disjunctive, and/or relational elements. Examples would be "Construct an office layout with no cabinets" or "Place on the west wall two cabinets with no other object between them" or "Place a cabinet on the left or right side of a desk but not behind or in front of it." Of course, it is difficult to represent such requirements, which express information of global or relational nature, in the form of properties of single units of a feature map. Logically, a feature map corresponds (at best) to a monadic first-order language, a language that only allows the use of one-place predicate symbols. With respect to descriptive power, monadic predicate logic is of course weaker than general (n-place or polyadic) predicate logic (it is, however, decidable, whereas first-order logic is only semidecidable, Habel, 1983, p. 123ff). Even if, in some cases, the internal handling of such complex sentences could be easily done on the basis of units representing sum variables (Paaß, 1992, p. 60f), it is unclear how to map them (supposing the user is allowed to formulate as many as he or she likes) onto fixed input vectors. Perhaps the only practicable solution in this, as well as in the context of ASM-based search, where the same problem arises, is to handle such aspects outside the associative network in a postprocessing mode on the basis of other tools like EXCEPT. Thus, once more, the need for a hybrid approach to the whole problem becomes evident.

Conclusion

Our basic research goal is to study and develop new methods for handling incomplete, vague, or contradictory knowledge. From a communicative or interactive point of view, which is relevant for the quality of the assisting modalities, it is important that application systems incorporating these methods be able to interpret inexact user instructions. As an example application field, we have studied the problems of processing imprecision during the construction and search of graphics. The user can incrementally specify the attributes of graphic objects (such as presentation graphics or diagrams of office furniture layouts). Adequate assistance consists here in completing and making more precise or consistent user specifications that are inexact in our sense. The techniques we are studying and applying are mainly based on nonclassical inference, such as nonmonotonic and associative reasoning, and on methods of planning under uncertainty.

Acknowledgment

As leader of the TASSO project, I am expected (among other things) to give overviews stressing the relationships between different research aspects that are worked out by other people. In this respect, I want to thank particularly my colleagues P. Henne, U. Junker, and G. Paaß, who have done the "real" scientific work addressed in this section.

2.4 EXPLANATION ABILITIES

Bernd S. Müller, Michael Sprenger

Explainability has a place among the assistance attributes: An assistant system can explain its functionality and behavior. During problem solving, at each point in the course from the problem statement to the presentation of the solution, explanations regarding what has been and explanations what will be and what could be are given on demand. Explanations concerning the domain knowledge, providing a static view on the domain, should be possible whenever needed by the user. The explanations may appear in different forms depending on the problem to be solved and the problem-solving step currently to be explained, for example, as problem-solving traces (reduced or full length), explanatory texts in natural language, graphics, animations, or video sequences.

There are several reasons why users may need explanations, such as to understand a solution better, for tutorial advice, or for verifying a knowledge base when developing systems. In general, the system's methods may differ from the methods of a human expert, and the plausibility of a solution and the insight into the solution process are limited in a complex system. Therefore, explanations may be not only useful for naive users, but for experts and knowledge engineers as well. Here, the boundary to help and tutorial systems is not well defined.

The explanations must be given at different levels of abstraction, just as the user wants to have them in a given dialogue situation and in relation to his or her individual needs. Different types of knowledge thus may be of great help for the successful performance of an explanation facility: knowledge that is explicitly structured into different levels of abstraction; knowledge about the user, dialogue preferences as prescribed by his or her domain experiences; knowledge about the actual dialogue situation; and knowledge about the domain enabling explicit explanations, motivations, and justifications for concepts, models and problem-solving methods.

With these different types of knowledge and the problem of their suitable representation as a guideline, work in the DIAMOD[13] project started. KADS

---

[13] DIAMOD (Natürlichsprachliche **DI**Alogstrategien für **MOD**ellbasierte Expertensysteme — Natural-Language Dialog Dependent Strategies for Model-Based

was felt to be a knowledge representation and acquisition paradigm that had enough of the necessary features to make it a promising candidate for detailed research. Although at the beginning of the DIAMOD project the structuring of knowledge into different layers of increasing abstraction made KADS especially attractive, in the final phase the reuse and genericity aspect of KADS also became important, leading to research plans for a generic explanation component.

Goals of the DIAMOD Project

The aim of the project DIAMOD (Horacek 1992a; Müller & Sprenger, 1991; Sprenger, 1993) was to develop explanation components for model-based expert systems that are based on the KADS approach[14]. Our testbed for implementing a first prototype was OFFICE-PLAN (Karbach & Voß, 1992a). The problem-solving methods behind OFFICE-PLAN were developed to find solutions for assignment problems in general. Our application appropriately assigned employees to rooms by taking into account a number of requirements. These requirements expressed several constraints for persons and rooms, for example, that group leaders and project leaders must be in rooms located near one another, or that smokers and nonsmokers should be in different rooms, and that a room had to provide several facilities needed by the employee(s) assigned to that room.

One of the hypotheses underlying the work in DIAMOD was that a model-based methodology for developing systems, like KADS, would facilitate the process of generating explanations. We decided to use OFFICE-PLAN as a testbed for our work on explanations, because it was implemented in the KADS-language MODEL-K (Karbach & Voß, 1992b).

As a second hypothesis we supposed that generating natural-language explanatory texts would improve the comprehensibility of explanations over canned text or formal descriptions (cf. Moore, 1985, pp. 21–39)). For example, natural language provides means for context sensitive expressions (e.g., use of pronouns); it is easier to create user adapted explanations; and finally,

Expert Systems) has been supported by the Ministerium für Wissenschaft und Forschung des Landes Nordrhein-Westfalen. Partners in the project were the University of Bielefeld (Dieter Metzing's group) and GMD Sankt Augustin (Thomas Christaller's group).

[14] DIAMOD concentrated on KADS I (Schreiber et al., 1993; Wielinga & Breuker, 1986; Wielinga et al., 1991). Future work would have to take into account Common KADS, as currently developed in the ESPRIT Project KADS II (e.g., Breuker & Van de Velde, 1994; de Hoog et al., 1992) and parallel developments like MoMo (Walther et al., 1992).

because of pragmatic features limited to natural language (e.g., conversational implicatures), explanations can be reduced to a small number of expressions without losing the required information. Reducing information relevant for an explanation was one of the central tasks in DIAMOD. Explanatory texts have to be short to be easy to understand for a user. This problem may be less severe for some of the current explanation components for one of the following reasons: First, the reasoning process itself may not be too complex to be presented as a whole. The line of reasoning may even be restructured and those elements of the reasoning process may be highlighted that are of interest for the user (Wick & Thompson, 1992). Second, many expert systems work interactively; that is, during the reasoning process they may ask the user for needed data. The user may also ask for explanations at these interaction points. The explanations generated then focus on the current part of the reasoning process, such as presenting rules or justifications for rules used by the system to perform the last step(s). Further parts of the reasoning process may be presented when asking follow-up questions.

In contrast, OFFICE-PLAN gets a description of the problem and then proposes possible solutions by using constraint mechanisms without further interaction with the user. The user can ask for explanations when the problem-solving process has finished. The most important problem when explaining the reasoning process (or part of it) is to find the relevant information: The original trace may consist of 20, 30, or even more pages. Thus, one main goal in DIAMOD was to reduce complexity of the information underlying the reasoning process. As a simple example, suppose the user asks, "Why is the person X in room 003?" This question can be answered with a verbalized trace that contains the complete information about the decisions responsible for the room assignment. But this would not only bore the user but probably confuse him as well, because of the amount of information the trace usually contains. In similar situations, humans not only give "correct" explanations, but also try to reduce the information to the relevant parts. To the question mentioned earlier, a human would perhaps reply something like "Group leaders must be in single rooms." The pieces of information left out here, such as that the person in question is in fact a group leader, are already known by the hearer or—as assumed by the speaker—could be easily inferred by the hearer. The problem of reducing the information provided by the system trace can be described by the following two substeps:

- Because of the amount of information used in a problem-solving process we need methods for selecting the relevant information for a specific explanation. This method has to take into account the specific user question, information on the actual problem-solving process, and the KADS model of expertise.

• The resulting set of relevant information has to be adapted to the actual situation given by a user and a dialogue model. This refers to the problem that the user not only needs relevant information with respect to his or her question, but concise texts, that is, text without information the user can easily infer (or even already knows).

Starting from this problem description, we can formulate DIAMOD's research interests in more detail by the following questions:

1. Is it possible to exploit the KADS model of expertise for selecting relevant information?

2. Are there special explanation techniques for constraint propagation methods?

3. Are there general methods for trace reduction, applicable not only on constraint propagation? Are theses methods valid only for synthesis (construction) domains?

4. How can we adopt relevant information to special user (and dialogue) needs?

5. When using KADS, is it possible to build generic explanation modules for interpretation models (i.e., for problem-solving methods, independent of a specific application or domain)?

6. How are different kinds of explanations combined?

7. Can we exploit a KADS model to find different levels of abstraction that can be used to filter information?

8. Which methods have to be used to structure information, so that coherent texts can be generated that are easy to comprehend?

In this section we will present some of the results of the project, especially referring to questions 1, 2, 3, and 4. For a first assessment of the KADS methodology for building explanation components (cf. Sprenger, 1992). A description of the influence of problem and methods used on the explanation facility is given in Sprenger and Wickler (1993). A short discussion of questions 6 and 7 can be found in Sprenger (1993). Because in DIAMOD the explanation structure mainly follows the structure found in the reduced trace, we have no final answer to question 8. A discussion of some aspects can be found in Horacek (1992a, 1992b).

Selected Results: Content Determination and Information Reduction

Explanations are in general reactions on specific questions that refer to a lack of knowledge. Therefore, strategies for generating different kinds of explanations have to be based on types of explanations that reflect the kinds of knowledge a question can refer to.

Terminological explanations. These explanations are concerned with the domain knowledge of a system, that is, the model of that part of the "real world" the system has to deal with. Typically this kind of knowledge consists of concepts, relations between concepts, instances of concepts, and domain-dependent rules. Techniques for generating this kind of explanation have already been established by several researchers working on generating natural-language texts from databases.[15] (McKeown, 1985)

Dynamic explanations. This is the "classical" type of explanations in the area of expert systems, of which the most important are initiated by *how* and *why* questions. It is called "dynamic", because the relevant knowledge used varies for every problem and because the flow of information during problem solving is examined. In most cases the expert system interacts with a user during the problem-solving process by asking him or her for specific information. At these points the user may ask *why* this information is needed or *how* the system has come to the conclusion (or part of a conclusion) just presented. In these cases "local" information is needed for the steps just performed. If the system is intended to give justifications for the reasoning steps done, additional information about the design decisions underlying th expert system is needed. But, in general, it would be helpful to have an explanation of the solution process as a whole. Such explanations have then to be produced on the basis of traces in which all information relevant for the problem-solving process is stored, such as facts and instantiated rules and tasks.

Explanations of generic problem-solving methods, without referring to a specific solution. The user may not only want to know how the system has come up with a specific solution to a given problem. He or she may also be interested in the problem-solving methods in general. This is

---

[15] For example, McKeown's TEXT system (McKeown, 1985) and Paris' work on exploiting a user model for text generation (Paris, 1989).

indicated by questions like "What is an abstraction process?", "How do you select the employee to assign next?" and "How are students assigned to offices?"

Explanations of meta-knowledge, for example, reflective knowledge of a system. Such explanations refer to the strategic or meta-knowledge of a system. In the REFLECT project (see chapter 2, section 2.1) the strategic knowledge is built by several reflective modules that have to decide how to proceed when the initial problem description is, for example, underspecified, redundant, or overcomplex. A typical question in the OFFICE-PLAN domain is, "How do you proceed when the problem is underspecified?"

For determining the content of an explanation, we have to look at the KADS model of expertise and its knowledge constructs, how they interact, and how they explicitly indicate the information they use during problem-solving processes. One advantage when using the model of expertise seems to be, that all the information on how domain concepts (and instances) are used in the inference process is provided and how the inference processes are controlled by higher level constructs (tasks) is declaratively represented. Determining strategies for choosing the content of natural language explanations involves fixing the different ways information is used and controlled in the system. In Fig. 2.19, the various relations between domain concepts and other pieces of expert knowledge that lead to different *explanation strategies* are outlined.

Here we give a brief overview of the results (for details cf. Sprenger, 1993). The strategies exploit the actual question type and the relations of the KADS constructs in the knowledge base and take into account information from user and discourse model. For all explanations it is possible to choose different levels of detail, depending on assumed user familiarity with the domain or the inference knowledge.

The two boxes in Fig. 2.19 represent two views of the expert system: The first box describes the *static knowledge*, that is, the knowledge base with its several layers and their constructs. The arrows denote the strategies that depend on the different relations between the constructs. They are called *strategies*, because with each relation one or more algorithms are given to select the content for an explanation describing the relation. Thus, the strategies can also be used to define a typology of explanations. For example, explanation strategy (1) denotes the terminological relation that holds between different concepts and relations (in KL-ONE-like terms: concepts

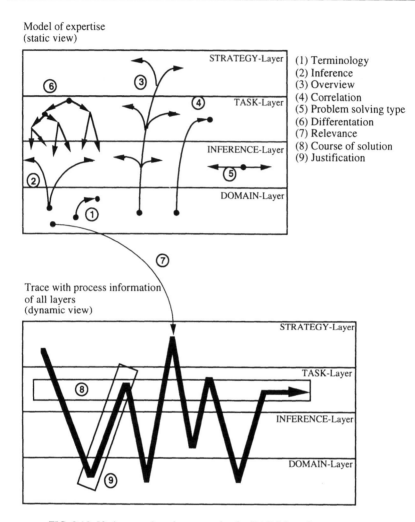

Model of expertise
(static view)

STRATEGY-Layer

TASK-Layer

INFERENCE-Layer

DOMAIN-Layer

(1) Terminology
(2) Inference
(3) Overview
(4) Correlation
(5) Problem solving type
(6) Differentation
(7) Relevance
(8) Course of solution
(9) Justification

Trace with process information
of all layers
(dynamic view)

STRATEGY-Layer

TASK-Layer

INFERENCE-Layer

DOMAIN-Layer

FIG. 2.19 Various explanation strategies for KADS-based expert systems.

and roles) at the domain layer. This strategy is used when the user wants a definition of a term or a comparison of two concepts.

Strategy (2) refers to the role a domain concept or relation may play at the inference layer, that is, how it is used in the various inference modules, for example, in an explanation for "How are students assigned to offices?" A peculiarity of the KADS approach is exploited in strategy (5): explanation of *domain-independent* problem-solving knowledge, which is rarely possible in other systems. In KADS all layers (with the exception of the domain layer) should be domain independent in the sense that they describe a general problem-solving method which could be adapted to other domains. For example, the upper layers of OFFICE-PLAN can be used to solve assignment problems in general; the assignment of employees to offices is only one possibility.

The second box illustrates the *dynamic view* of the expert system: The curve crossing the four layers denotes the trace of a single problem-solving process. The strategies in this box denote (at least) three different kinds of explanations that are concerned with dynamic knowledge, that is, explanations which provide the user with information about a single problem solution (e.g., explanations for questions like "Why is person X in Room 001 and not in Room 005?"). Although strategies (8) and (9) are concerned with highlighting interesting elements of a solution or justifying a solution (or part of it), in (7) the *relevance* of a domain concept is explained; that is, the role a concept plays in a particular problem-solving process is described.

An explanation of how a problem has been solved can be given for a complete solution, for a partial solution, or for a part of a complete or partial solution. The first kind of explanation seems to be especially difficult because the enormous amount of information in a complete solution trace is far too detailed to be an informative explanation for the user. Thus, the trace information must be reduced to the most relevant points.

Three different kinds of knowledge can be exploited for doing the information reduction job. First, there are pure logical criteria. The trace can be viewed as a tree of single inference steps; thus, not all parts of the trace directly belong to the inference chain of the part of the solution in question. Second, the content of the single steps can be analyzed with regard to their relevance for the solution. Finally, there are specific characteristics of natural language that can be used to avoid overinformative texts. We next present the main ideas behind these techniques.

The first of these techniques depends on the problem-solving method used in DIAMOD, a constraint propagation mechanism. The trace as the starting point for dynamic explanations contains the relevant information in terms of constraints evaluated and the effects imposed on the assignments. One employee is selected and all constraints relevant for this employee are applied. At the end of such a local propagation step the system makes assumptions with respect to the assignments made so far, and it comes to a solution (or a set of possible solutions) or it discovers an inconsistency. These steps are applied until all employees are assigned to offices. In case of dynamic questions such as "Why has employee X not been put into Room 2?" or "Why must employee B be put into Room 3?" a strategy is chosen based on an idea in the resolution principle. In general, questions of the type "Why is ⟨fact⟩ (not) valid?" can be answered by making an *indirect proof*. As the ⟨fact⟩ in the question is in contradiction with the produced solution(s), it is hypothetically added to the knowledge base, and the constraint mechanism again is used to deduce a contradiction.[16] The trace of this second application of the constraint mechanism is then finally used to produce an answer to the original question. There are two main advantages with this strategy: First, it turned out that the second trace is (in most cases) much shorter then the trace of the original solution, because it is tailored directly to the question. Second, by this deduction the relevant information for answering the question is better structured to produce a comprehensible explanatory text. This procedure is then followed by two reduction steps, called linear trace reduction and reducing backtracking, which reduce the inference steps to those that lead to the contradiction found, and which reduces identical proof steps on each level to just one. These reduction steps are applicable to all traces with a tree structure (for details see Wickler & Heider, 1992).

The following simple example demonstrates how the content for an explanation is selected and how the information is reduced to the relevant propositions. For simplicity, we include in our example only four employees (A, B, C, and D) and four rooms (1, 2, 3, 4), where the rooms with numbers differing by one are next to each other. After the assignment is made, the user asks, "Why cannot C be assigned to 1?" For answering this question, the system assumes the constraint in the question (C must be in room 1) and tries to proof it. The resulting trace passed to the component for information reduction is shown in Table 2.1.

The information reduction component reduces the following propositions (for details see Wickler & Heider, 1992): Steps 10 and 9 are redundant, because the resulting conflict is already established in step 7. Thus, step 8 can also be left out, because it introduces the assumption for steps 9 and 10. Finally, steps 3 to 6 are not relevant for answering the question, because they do not affect the possible assignments for D. The resulting propositions are 1, 2, and 7, as shown in Table 2.2.

---

[16] For those cases, where ⟨fact⟩ does not contradict the current knowledge base, the question can be reformulated so that an indirect proof can be applied. For details cf. Wickler & Heider (1992).

TABLE 2.1

Step	Employees involved	Constraints	Possible assignments: employees, rooms			
			A	B	C	D
1	C	C in room 1 (*assumption*)			1	
2	C, D	D in a room with SUN			1	1
3	B, D	B and D in different rooms		2,3,4	1	1
4	A, D	D and A in different rooms	2,3,4	2,3,4	1	1
5	A, C	C near to A's room	2,3	2,3,4	1	1
		Assumption: A in room 2				
6	A, B	B near to A's room	2	3	1	1
7	C, D	rooms should provide space	C and D in conflict in room 1			
		Assumption: A in room 3				
8	A, B	B near to A's room	3	2,4	1	1
		Assumption: B in room 2				
9	C, D	Rooms should provide space	C and D in conflict in room 1			
		Assumption: B in room 4				
10	C, D	Rooms should provide space	C and D in conflict in room 1			
		No assumption left,				
		refutation proved				

TABLE 2.2

Step	Employees involved	Constraints	Possible assignments: employees, rooms			
			A	B	C	D
1	C	C in room 1 (*assumption*)			1	
2	C, D	D in a room with SUN			1	1
7	C, D	Rooms should provide space	C and D in conflict in room 1			
		No assumption left,				
		refutation proved				

The resulting explanation is: "D must be in room 1, because he needs a SUN computer. But room 1 does not provide enough space for D and C".

Another procedure for reducing the amount of information in the solution trace is presented in Horacek (1991b). It takes into account the relevance of single inference steps and the adequacy of the presentation structure. The mechanism contains four main steps. First, a set of potentially relevant requirements is determined out of the complete trace, depending on the fact in the user's question. For example, if the user asks why the assigment for employee X is valid, only the inference steps potentially relevant until X is assigned are of interest, but not any information concerning persons assigned later. Second, the set of requirements is prestructured into several partitions, according to their influence on aspects relevant to the user's question. In our example, the partitions are determined by every change in the possible solutions concerning X's assigments. Third, the set of requirements in each partition is reduced, so that only those remain that directly affect the solution. For example, all requirements can be left out that have effects on other employees but not on X. Finally, a decision is made concerning the degree of complexity of the explanation; ithat is, in case there is still a large set of requirements, it is not necessary that all partitions are presented (cf. Horacek(1991b).

Finally, natural language has several characteristics that can be used for further reduction. For example, phenomena as grouping ("x, y, and z are green" instead of "x is green, y is green, and z is green") or elliptification are well known to present condensed texts. Horacek (1991a) has developed a method for generating concise texts, taking into account the Gricean principle of relevance (Grice, 1975) to simulate the phenomenon of *conversational implicature*. For example, if you ask your colleague "Are Jack and Jill still in room 113?", and your colleague answers with "No, Jack is now in 115", you can be sure that Jill is still in 113, although not explicitly mentioned in the answer, and not logically inferable. Otherwise your colleague would have told you (or at least have added something like "I don't know about Jill"). Conversational implicatures are information that we easily infer in everyday discourse. In Horacek's model, a set of contextually motivated inference rules is applied in an anticipation feedback loop to reduce the set of propositions in an explanation to a subset that, in the actual dialogue context, can still be correctly interpreted by the user. Thus, concise explanations can be generated, that is, explanations without explicit information the user already knows or can easily infer.

The DIAMOD approach resulted in the implementation of a prototype system that generates natural language explanations in German for a restricted set of questions in the OFFICE-PLAN system. The main question types are terminological questions concerning concepts, relations, and instances (e.g., "Who is a project leader?"); questions about methods (e.g., "How do you select the next employee?"); and questions concerning a problem solution (e.g., "Why must P assigned to Room 2?"or "Why cannot A and P be assigned to one office?"). The prototype system includes the described information reduction techniques of Wickler and Heider (1992) and of Horacek (1991a). The result of content selection and information reduction is a functional structure that is passed to the component responsible for generating the surface structure (Peters, 1993).

Further Results of DIAMOD and Future Work

Not less important, especially for future work on explanation facilities, are some of the theoretical findings of DIAMOD that did not become part of the prototype explanation component.

Becker (1993a, 1993b) looked into user modeling in the context of explanation and KADS-like modeling with emphasis on:

1. Design and implementation of a generic user modeling system for the acquisition and non-monotonic maintenance of propositional user models in dialogue systems.

2. Design of a method for detecting the distribution of domain knowledge in a user population; the method can be used as a basis for judging about an individual user's global knowledge level.

3. Methods seeking to exploit the identifiability of functionally coherent subdomains of KADS-based systems for user modeling purposes.

Müller (1991) analyzed some existing natural-language generation systems and compared them according to their flexibility and coverage: highly flexible natural-language generation systems seem still to be beyond the state-of-the-art. Müller (1992) made a first attempt to compare the classical rhetoric with recent text generation methods. For architectural considerations, for the design of the pragmatic phase of the generation process and because of the overwhelmingly rich description of rhetorical language phenomena, the results of classical rhetoric should be analyzed in depth. Meier (1992) proposed a semantic model for the content selection phase of the generation process. Müller and Becker (1991) discussed fundamental problems in knowledge representation for natural language generation by using examples of the OFFICE-PLAN domain: The need for augmenting the representational power of existing representation methods still is great, as the discussion about symbol grounding and the search for nonsymbolic representation methods in the last decade have shown. Old and new paradigms are not yet reconciled or successfully combined. There are no signs of a major breakthrough in the near future.

During the workshop "Explanation Facilities for Model-Based Expert Systems" (Workshop Explanation, 1992) first steps were made towards a comparison of KADS and the Explainable Expert Systems paradigm of the Swartout research group (cf. Swartout, 1992)) with special emphasis on their explanation capabilities (Wielinga & Swartout, 1992; see also Meier, 1991).

Decreasing interest in knowledge engineering may impede research in explanation that will nevertheless be a very relevant aspect of an intelligent use of computers for quite a time.

Conclusion

The explanation ability is one of the basic properties a knowledge-based system implementing assistant behavior must have. After positioning explainability into the context of the assisting computer paradigm, experiences from the project DIAMOD have been reported, going into some detail with regard to KADS, the knowledge representation and acquisition technique used there, and its advantages for explanation purposes. As major results of DIAMOD we briefly described the explanation strategies for KADS models, the trace reduction algorithms, and the exploitation of conversational implicature for text reduction purposes. Finally, we related some of the findings in DIAMOD to future work in explanation.

3 Systems with Domain Competence

3.1 KNOWLEDGE ACQUISITION ASSISTANCE

Stefan Wrobel

One of the central goals of the Assisting Computer (AC) research program carried out at GMD's Institute of Applied Information Technology (FIT) over recent years (see section 1) was to break ground for computer systems that realized a new division of labor between human and machine. Where in the past the goal of "data processing" was to build systems that eliminated operator intervention as much as possible, the AC research program was based on the insight that for more challenging tasks, it is impossible to succeed without the knowledge and intelligence available to the human user, and computers systems should therefore be "assistants" that cooperate with their users instead of replacing them.

To reach this goal, work in the AC program has identified a number of properties or capabilities that computerized assistants should possess so they can effectively cooperate with humans (see Hoschka, 1991, and section 1 of this book). In this section, we are particularly concerned with two of these assistance properties, namely, *domain competence* and *learning and adaptivity*. The goal of domain competence reflects the fact that assistance support is never independent of the particular problem that is being addressed. If an assistance system can be equipped with special knowledge about a domain or customized for a particular kind of task, it will be able to offer a much higher level of support. Because the task of acquiring this kind of specialized knowledge and representing it in the computer (known as knowledge acquisition) is very difficult to perform manually without additional assistance, one important subgoal of the AC program was to examine the possibilities of building assistance systems specialized toward this knowledge acquisition task.

One way of building up knowledge, of course, is to learn it from examples or other available data, which brings in our second assistance topic, learning and adaptivity. When the AC program was started, the field of machine learning (ML) had already identified a large number of techniques that were capable of for example, taking examples as input and inducing from them general rules or decision trees. It was much less obvious at the time, and still is an important problem today, how these techniques are to be properly integrated into assistance systems so they can really effectively cooperate with a user—by themselves, most of the basic ML algorithms follow the traditional data processing paradigm, taking data as input, processing them, and producing results as output. What role should these techniques take in the knowledge acquisition process and in an assistance system?

This research question and its associated subproblems were researched and are still being researched at GMD in the ESPRIT projects MLT (Machine Learning Toolbox, 1989-1993), ILP (Inductive Logic Programming, 1992-1995) and ILP2 (Inductive Logic Programming 2, since January 1996).[1] In this section, we present an overview of the main results of this research and describe its main future directions. The section is organized as follows. We first discuss our view of the general nature of the knowledge acquisition process as it is described in the balanced cooperative modeling paradigm (Morik, 1993) and identify a number of specific properties that assistant systems in this paradigm should possess. We then introduce the MOBAL system, a full-featured knowledge acquisition and machine learning system that instantiates the general balanced cooperative modeling architecture.[2] The following subsection is devoted to a description of the various assistance services offered by MOBAL, including the automatic acquisition of first-order rules from examples and support for the revision of knowledge bases. We finally discuss how these assistance capabilities of the MOBAL system

---

[1] The work described in this section was partially supported by the CEC's ESPRIT program under contract no. 6020 ("Inductive Logic Programming"). Previous development of MOBAL was partially supported by ESPRIT contract 2154 ("Machine Learning Toolbox").

[2] GMD grants a cost-free license to use MOBAL for academic noncommercial purposes. The latest release of MOBAL can always be accessed via the Web pages of the ML group (http://nathan.gmd.de/projects/ml/home.html) or directly via FTP (host ftp.gmd.de, directory /gmd/mlt/Mobal).

can be embedded into other application systems, a topic that will be more and more important in the future ("embedded adaptivity").

Balanced Cooperative Modeling

In her work on the nature of knowledge acquisition processes, Katharina Morik (1989, 1991) has introduced the notion of knowledge acquisition as a cyclical *sloppy modeling* process: Starting from an initial domain model that out of necessity is incomplete, inconsistent or incorrect ("sloppy"), the user makes additions or revisions in order to improve the model, observes and evaluates the effects of these changes, makes further modifications, and so forth, until the overall quality of the model is judged satisfactory. This iterative process of knowledge acquisition occurs independent of the kind of system support that is available, but depending on how machine learning is used in the assistance system, very different interaction styles, and very different levels of support, arise.

In Morik (1993), three different cooperation styles were distinguished:

- One-shot learning. This scenario corresponds to the classical "data-processing" approach: The learning system is used in a compiler-like fashion to induce rules from background knowledge and examples.

- Interactive learning. In this form of cooperation, the learning system controls the acquisition process by asking questions to the user. CLINT (De Raedt & Bruynooghe, 1992) and MIS (Shapiro, 1983) are (first-order) ML systems of this type.

- Balanced interaction. Both user and system can contribute to the evolving domain model in a mixed-initiative fashion; the system supports not only addition but also inspection, revision, and restructuring of the model.

As argued in Morik (1993), the balanced style of interaction (also referred to as *balanced cooperative modeling*), has a number of advantages compared to the other two in supporting knowledge acquisition by learning. In contrast to purely query-driven learning systems ("interactive learning"), the user controls the acquisition process and can work on whatever part of the model he or she sees fit. In contrast to batch learning systems, however, the system still provides feedback interactively, complements the user's inputs, and points to problems (e.g., inconsistencies) in the emerging model. As in query-driven learning systems, and unlike one-shot learning, all information need not be present right away. Unlike in query-driven systems, however, the user can provide whatever information is available at whatever moment. Most importantly, perhaps, the user can revise the emerging model at any point and does not have to go through a query process again, nonetheless, the revision step is supported by the system, which is not the case in batch learning systems.

From the general goals of sloppy modeling and balanced cooperative modeling, one can derive a number of specific properties that must be met by an assistant system for the construction of domain models (Wrobel, 1988):

Flexibility. Domain modeling is a task that is characterized by many sudden ideas about structure and content of the model and frequent changes in the focus of attention. To support this, the user/system dialog must be mixed-initiative, so that the system can follow the user instead of locking him or her into a fixed interaction order.

Reversability. At any point during modeling, it may become necessary to change the content or structure of the model in a nonmonotonic fashion. The assistant system thus must offer reason maintenance and revision operations that allow previous modeling decision to be easily reversed.

Integrity and consistency maintenance. An assistant system should free the user from housekeeping chores so that he or she may concentrate on the difficult parts of the task. Thus, the system should automatically keep track of inconsistencies or integrity violations in the emerging model.

Liveliness. The user must receive constant feedback on the properties of the model that is being constructed. After every change to the model, changes should be propagated so that their effects can be evaluated by the user.

Inspectability. Closely related to liveliness, an assistant system for knowledge acquisition must place an emphasis on offering intuitive, graphical means of inspecting the available knowledge. This is especially true for the construction of large domain models, where textual representations can become very difficult to handle.

Transitionality. There should not be a boundary between user actions and system actions. The assistance components of the system and the user should both be working on the same knowledge sources so that manual and automatic acquisitions can happen at the same time and in cooperation.

In order to implement the balanced cooperative modeling paradigm and the design goals which it entails in an assistant system for knowledge acquisition, a number of choices must be made with respect to knowledge representation, user interface, and the available support services. In the next section, as an example of a balanced cooperative modeling system, we will see how these choices were made in the MOBAL system.

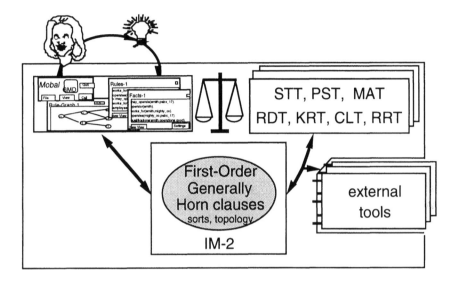

FIG. 3.1 MOBAL's system architecture.

MOBAL

MOBAL is a multistrategy learning system (Michalski, 1993) that consists of a collection of cooperating learning modules organized around a knowledge representation subsystem (Figure 3.1). The knowledge representation subsystem, the inference engine IM-2 (Emde, 1989, 1991; Morik et al., 1993), is responsible for manipulating the entries that make up a domain model or knowledge base (see later discussion), for making inferences with these entries, and for offering reason maintenance services so that inferences are automatically retracted when they become invalid. As knowledge representation, MOBAL uses a first-order function-free Generally Horn logic, augmented with several other knowledge sources. We briefly explain the different types of representation items that are available (more detail can be found in Morik et al., 1993). All examples are taken from a telecommunications application of MOBAL (Sommer et al., 1994).

A *fact* is simply a predicate symbol with an appropriate number of constants as arguments, as in

    may_operate(bode,pabx_17) .

Facts are thus required to be ground (contain no variables), and function symbols other than constants generally are not used in MOBAL's representation (function-free). In contrast to many other ILP systems that use the standard two-valued Prolog semantics with negation as failure, MOBAL employs a four-valued paraconsistent semantics (Morik et al., 1993) that allows negation and contradictions to be explicitly represented with truth values true, false (indicated by not), both (contradictory) and unknown (usually not listed in the knowledge base). Examples of false and contradictory facts are

    not(may_operate(meyer,pabx_17)) .

    both(may_operate(bode,pabx_15)) .

If desired, it is possible (but not required) in MOBAL to declare the argument sorts of predicates:

    may_operate/2: <employee >, <system >.

This declares may_operate as a binary relation taking employees and systems as arguments. MOBAL's sort taxonomy tool STT can use this information to create a *sort lattice,* but will also automatically generate such declarations when they are missing (see next subsection). Additional information about predicates can be expressed in the domain topology. A topology consists of named groups of predicates that can be linked to each other. For example,

    'Access Rights' -

    [may_operate,may_access] arrow ['Employees','Qualifications']

defines 'Access Rights' as a topology node consisting of the predicates may_operate and may_access, the computation of which is intended to depend on the predicates expressing information about 'Employees' and those expressing information about 'Qualifications'. Figure 3.2 shows an example of a topology.

A *rule* is a clause that consists of a number of premise literals and a single conclusion literal, as in

    works_for(P,C1) & operates(C2,S) & subsidiary(C1,C2) & technical(P) ->
    may_operate(P,S) .

As in Prolog, variables are written uppercase in rules and are assumed to be universally quantified. Just as facts, the premises or conclusion of a rule can

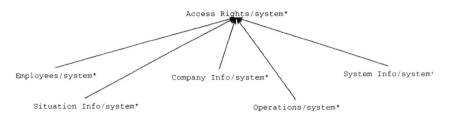

FIG. 3.2 Example of a predicate topology.

have any of the four possible truth values. MOBAL's representation thus closely corresponds to generally Horn programs (Blair & Subrahmanian, 1989). Additional predefined predicates can be used to perform arithmetics (e.g., add) or to draw inferences based on the current state of the knowledge base (e.g., \+ to denote currently unprovable facts) (Emde, 1989; Morik et al., 1993).

Rules are used by the inference engine for forward and backward inferences, whereas *integrity constraints* are not used for inferences but are simply checked on the current knowledge base. They can therefore take a more general form than rules, and can be used to state disjunctive or negative information that cannot be expressed using rules. The integrity constraint

operator(X) & manager(X) ~> .

states that no one can be both manager and operator, and

employee(X) ~>operator(X) ; manager(X) .

states that every employee must be either operator or manager.

In addition to facts and rules themselves, MOBAL also offers higher-order constructs called metapredicates, metafacts and metarules that can be used to declaratively state properties of predicates and inferential relationships between rules.[3] For example, given the metapredicate

inclusive_1(P,Q): P(X) -> Q(X).

we can use the metafact

inclusive_1(operator,employee) .

to state that all operators are employees. Internally, this metafact is translated into the rule

---

[3] It is also possible to use metametapredicates and metametafacts to describe the properties of metapredicates.

operator(X) -> employee(X) .

The transitivity of inclusive_1 can be expressed with a metarule:

inclusive_1(P,Q) & inclusive_1(Q,R) -> inclusive_1(P,R) .

For all these representational items, MOBAL's graphical knowledge acquisition environment can be used to view textual and graphical presentations (see Fig. 3.3), and to make addition, modifications, or deletions. The inference engine is responsible for making forward and backward inferences with the knowledge base, and thus provides feedback about the operational properties of the emerging model. MOBAL's learning tools, finally, provide additional services and cooperate with the user to complete and improve the emerging domain model.

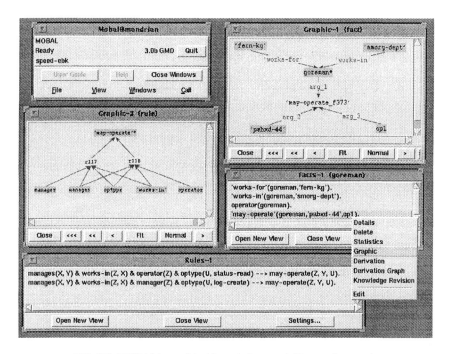

FIG. 3.3 MOBAL's graphical knowledge acquisition environment.

MOBAL's Assistance Services

The central principle of the balanced cooperative modeling paradigm, and thus the central idea behind MOBAL's assistance services, is the idea of *balanced* cooperation, that is, for every kind of knowledge source that can be entered by the user, there should be system components that are capable of working on this kind of knowledge and if possible, capable of automatically generating and revising this kind of knowledge (see Morik, 1993). Clearly there will always be a qualitative difference between knowledge input by the user and knowledge generated automatically, because only the user has access to the external world to validate the model, whereas the machine must rely on user inputs alone. The role of the system is to make implicit knowledge explicit, to perform tasks that are tedious or too time-consuming to do manually, whereas the user employs the machine's results as feedback about necessary changes to the domain model.

Inference and Structuring Services. In MOBAL, almost all knowledge sources can be filled and revised both manually and automatically. At the basic level of *facts*, the inference engine IM-2 (Emde, 1989) is capable of using rules to infer additional facts from the user's inputs. Whenever the user changes the set of input facts, the set of derived facts is automatically changed by removing inferences that are no longer true or adding those that have become valid (reason maintenance).

For information about predicates and their argument sorts, two modules are available to construct both *sort lattice* and *predicate topology*. The *sort taxonomy tool* STT (Kietz, 1988) records the actual usage of terms at the different argument positions of a predicate in input facts, and dynamically constructs and permanently updates a lattice of so-called *argument sorts.* Whereever necessary, predicate declarations are added automatically, and whereever declarations have been entered manually, their corresponding sorts are inserted in the proper place in the lattice (Fig. 3.4). For the automatic construction of a topology, the *predicate structuring tool* PST (Klingspor, 1991) analyzes the set of rules input by the user, builds a call graph among them, and employs several graph abstraction operations to build a topology that the user can then refine further.

Rule Induction Services. For the automatic acquisition of rules themselves, the *rule discovery tool* RDT (Kietz & Wrobel, 1992) can be used and performs a classical ILP learning task (akin to the descriptive ILP setting of Helft, 1989, or De Raedt & Lavrac, 1993). Given a knowledge base that contains facts about a specified target predicate and additional facts and rules as background knowledge, RDT finds all most general rules which use the target predicate as conclusion and are true for the given examples (meet a user-specified acceptance criterion). To constrain the search, metapredicates

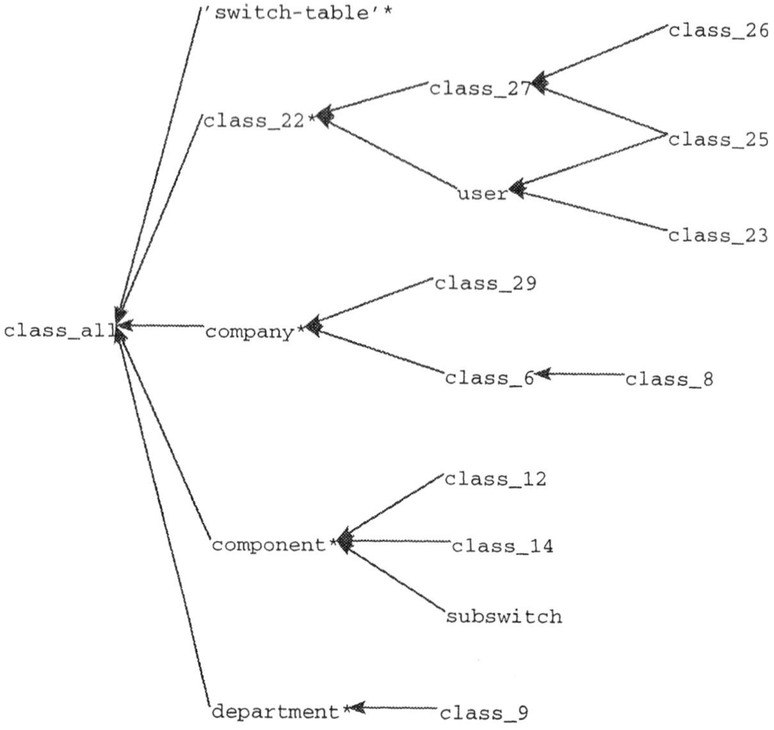

```
          /'switch-table'*
                                              /class_26
                          /class_27
          /class_22*                          /class_25
                          user
                                              class_23

                          /class_29
class_all    company*
                          class_6    class_8

                          /class_12
          component*       class_14

                          subswitch

          department*    class_9
```

FIG. 3.4 Example of a sort lattice.

are given as *rule schemata,* and RDT will search only for instantiations of the given rule schemata. Using RDT as a subroutine, the *concept learning tool* CLT (Wrobel, 1994a, 1994b) can search for all rules about a target predicate (premise or conclusion), and is capable of evaluating the quality of the resulting set of rules with respect to its suitability as a concept definition.

The acquisition of rule schemata is supported both by the *model acquisition tool* MAT (Thieme, 1989) and the learning algorithm INCY (Sommer, 1993). When given a rule as input, MAT is capable of automatically abstracting a metapredicate definition from it. INCY, on the other hand, proposes new metapredicates specifically as rule schemata for learning: In a bottom-up learning pass, potential rules built according to a heuristic criterion are tested against the knowledge base, and the promising ones are entered as metapredicates and rule schemata for RDT. For integrity constraints, RDT has been extended into ICDT, the integrity constraint discovery tool (Englert, 1995).

In its most recent release, MOBAL has also been augmented by an "external tool" facility (Emde et al., 1993). An external tool is a learning system developed outside of MOBAL, but interfaced to the system using special interface modules. As soon as such an interface module is available, the external tool (after it has been obtained from its authors) can be "loaded" into MOBAL, and after this, can be used in the same fashion that the internal module RDT was used before. When an external tool is to be run, the interface routines produce input files in the appropriate format, run the external tool with the right parameters, read its output file and enter the resulting rules into MOBAL's inference engine. Parameters and other knowledge sources for external tools are declared in the interface file and are then managed by MOBAL. At present, interface files are available, for example, for FOIL (Quinlan, 1990), MFOIL (Dzeroski & Bratko, 1992), GOLEM (Muggleton & Feng, 1990), GRDT (Klingspor, 1994) (a grammar-based version of RDT), and RDT/DB (Lindner, 1994), which is coupled to a relational database system and can be used to perform *knowledge discovery in databases.*

In addition to these learning tools from external groups, we are also using the external tool facility to integrate newly developed algorithms of our own. COLA (Emde, 1994) is an ILP learning system that learns from very few examples, employing the first-order clustering system KBG (Bisson, 1992) as a subroutine. The INCY algorithm (Sommer, 1993) mentioned above, and its successor LINK (Sommer, 1994a), are also integrated as external tools.

Revision and restructuring services. Two additional modules are available for aiding the user in making changes to the existing model. The *knowledge revision tool* KRT (Wrobel, 1993, 1994b) is an interactive module with its primary purpose to act as an assistant to a user who wants to remove one or more incorrect inferences (KRT is capable of performing multiple simultaneous revisions) from the knowledge base, that is, a user who wants to fix incorrect entries. Roughly speaking, the services of KRT are the following. First, KRT identifies, from the information stored in the inference engine, exactly which entries in the knowledge base participated in the derivation of the incorrect inference. This derivation information is then graphically presented to the user. Second, KRT computes the set of all possible minimal knowledge base revisions that would be necessary and sufficient for removing the offending inferences, based on the theoretical concept of a *minimal base revision* (Wrobel, 1993). Because the set of possible revisions can be large, the third service offered by KRT is a heuristic proposal of which of the possible minimal revisions to choose, based on a two-tiered confidence model (Wrobel, 1994b). When the user has selected the proposed or some other revision, KRT performs the necessary changes to the knowledge base to implement the revision and, as a fourth service, offers to reformulate rules by trying to find existing or new concepts that could be as additional premises.

The *ruleset restructuring tool* RRT (Sommer, 1994b), on the other hand, is used whenever a knowledge base, even though correct, is nonoptimal in terms of performance or understandability. RRT offers a number of operations for restructuring such rule bases, including a rule base stratification operator. This operator, related to the predicate invention operators known from inverse resolution work (such as intraconstruction), analyzes the premises of the selected set of rules, and heuristically detects common patterns in them. These common patterns, referred to as "common partial premises", are used to define a new predicate which then replaces all occurrences of the common premises. Such an operator thus allows the user to shorten long rule premise lists by identifying commonalities; this can greatly enhance the understandability of the knowledge base.

Towards Embedded Adaptivity

Given that a significant amount of assistance is thus available during the construction phase of a knowledge base, an interesting question is of course whether these capabilities can also be made available during the usage phase, that is, in the deployed application system that uses the model that was built up. In cases where the application problem is a relatively isolated task that can be solved almost exclusively with a knowledge-based system, one could simply take the developed model and incorporate it into the knowledge base of a shell that is then used to solve the application problem. Very few applications, however, can be solved using a knowledge-based system alone; more often, the knowledge-based part is responsible for only a small portion of the functions of a larger conventional application software system.

As an example, consider the telecommunications application that we have been using in this section to illustrate our knowledge representation (Sommer et al., 1994). This application, developed by Alcatel Alsthom Recherche, deals with the problem of access control in telecommunications networks. One standard solution to this problem is to use a database of manually assigned access rights against which each access is checked by the operating software of the switching systems. The disadvantage of this approach is that a given policy of assigning rights can be enforced only by organizational, not technical, means. In the MLT project, a prototype was constructed using MOBAL to show that explicit logical rules could be used to infer access rights from information about employees and systems, thus ensuring a consistent policy. Using the learning module RDT, a sample database was analyzed to derive a first set of rules, which could then be executed by the inference engine or revised using KRT.

If these services are to be used by the security manager in deployed system, one cannot make do with the standard interface offered by MOBAL as described earlier. Even though the services that are used, like inference, can still be the same, the interface must be kept entirely in terms of the application domain. Furthermore, the operating or access control software of the switching system must be capable of accessing the knowledge base to enquire about access rights, replacing the database queries. To realize both of these requirements, it must be possible to *embed* the capabilities of MOBAL into a custom interface and to couple them to other software.

In MOBAL, we are therefore presently supporting two possible approaches toward such *embedded adaptivity*. The first is a traditional software engineering approach. Due to the modular software architecture of MOBAL, it is possible to integrate one or more of its modules into an application system where it would realize adaptive capabilities. This approach requires modules written in different programming languages and with different goals to be integrated, however, so it is suitable only for cases in which complex and very deep integration is aimed at.

In other applications, we can imagine that the application software needs only very limited "services" from the learning system. This suggests that for integration in these cases, we can benefit from using a client/server-oriented conceptualization, where the learning system is a "server" offering knowledge-based inference and learning services, and the application system is the "client" requesting these services. In the new release of MOBAL (3.0), we have therefore chosen to base the interaction between MOBAL's system core and its user interface on a network-based client/server protocol. MOBAL's user interface (built with Tcl/Tk) is now a client that requests information from MOBAL's core through a TCP network stream protocol. Similarly, for example, in an applications such as the one just described, a custom interface can interact with the system core without ever showing this to the user, or a conventional software system can send requests to use, for example, the inferencing capabilities of MOBAL.

This kind of client/server integration offers application developers a chance to incorporate learning capabilities into their systems without the need to have access to or to understand the internal code of the learning system. All that is required is an understanding of the communication protocol understood by the learning system, and the ability to write to and read from TCP streams. Application and learning system can therefore be developed in different programming languages, on different platforms, and can even run on different computers when they are used to improve performance. Because the required communication bandwith is small, even remote access from small systems is possible.

Conclusion

In this section, we have seen how the MOBAL system supports the user in the process of building up a domain model and using it in an application system . Let us now briefly return to the design goals put forth at the beginning of this section and see how they are supported in MOBAL. For the goal of flexibility, due to its user interface based on individual editing changes to the model, MOBAL is flexible enough not to constrain the user's actions; every piece of knowledge can be input at every moment. As for liveliness, MOBAL automatically computes inferences and updates its sort taxonomy after every change of the knowledge base to provide immediate feedback. Inspectability is supported through a graphical user interface. With respect to transitionality, as required by balanced cooperative modeling, both user and system can work on the emerging model, so there is no transition between different modes. Finally, reversability is supported by the inference engine's reason maintenance, the knowledge revision tool and the ruleset restructuring tool.

For the future, more work is of course possible and necessary to further improve MOBAL's services as a knowledge acquisition and embedded adaptivity assistant. We can only mention a few of these points. We already noted that more learning algorithms are currently being added to the system. Improvements in the basic knowledge representation formalism could include better support for the notion of cases and of hierarchical structuring to more closely match users' familiar concepts. Furthermore, as experience with our users shows, there is room for improvement in the graphical user interface whose available focusing facilities could be improved to more easily show a user what he or she needs to see at a particular point. The restructuring modules, besides being controlled manually, ideally would have a built-in notion of understandability so they could automatically propose appropriate changes. For embedding into other applications, the current network-based protocol could be extended in the direction of standard sharing and interoperability formats such as KQML/KIF or CORBA.

Acknowledgments

MOBAL has been developed by Werner Emde, Jörg-Uwe Kietz, Katharina Morik, Edgar Sommer, and the author with contributions from students Roman Englert, Volker Klingspor, and Marcus Lübbe. The X-Windows interface of the current release (3.0) was developed using Tcl/Tk with the help of Sven Delmas (TU Berlin).

3.2 ASSISTANT FOR KNOWLEDGE DISCOVERY IN DATA

Willi Klösgen

The rapid growth in number and size of databases requires new methods and systems supporting the partial automation of data exploration. Hence, there is increasing interest in tools that can assist analysts in "mining" in large amounts of data. The new research area *knowledge discovery in databases* has been established around these general goals in the last years. Today, discovery research has already matured worldwide and is being transferred to applications. The most successful applications appeared in the areas of greatest need, where the databases were so large that a "manual" analysis supported by query languages and statistical packages was highly insufficient.

The features of database query systems and statistical packages are convenient for "manual" data analysis, supporting the test of a single hypothesis, but not assisting the generation and evaluation of a space of hypotheses. Query languages cannot satisfactorily represent verification queries requiring the computation of relative frequencies and statistical hypotheses. Statistical packages mainly support the test of an individual hypothesis specified by the user, not the search for findings. They produce general statistical indices and no conceptual interpretations.

In "Gold Mine of Data," *Business Week* (March 21, 1994) reported on how companies use their data bases for improving service. Competitive pressure forces companies to probe the gold mine of data they have been gathering for so long. Massively parallel computers are used to seek out "faint but significant" patterns. Companies including AT&T, ConRail, and Otis all report significant results in improving service.

The article "Firms Set to Mine for Supercomputing Gold" in *ComputerWorld* (January 17, 1994) reported that some of the U.S. top retailing, financial, and transportation companies are interested in data mining, using massively parallel processing systems. The article quoted an IBM manager saying that "Many commercial users ... will soon be getting a terabyte of information a day."

It is estimated that the U.S. market for data mining today is about $500 million. Data mining or discovery in databases is the search for relationships and global patterns that exist in large databases but are "hidden" among the vast amounts of data, such as a relationship between patient data and their medical diagnoses. These relationships represent valuable knowledge, if the database is a faithful mirror of the domain.

One of the main problems for data mining is that the number of possible relationships is very large, thus prohibiting the search for the correct ones by simply validating each of them. Hence, we need intelligent search strategies. Another important problem is that information in data is often noisy or missing. Therefore, statistical techniques should be applied to estimate the reliability of the discovered relationships.

Discovery systems support an analyst in performing knowledge discovery processes with the aim to find out new knowledge about an application domain. A discovery process cannot be specified in advance and automated completely, because it depends on dynamic, result-dependent goals of the analyst and emerges iteratively. Typically, a process consists of many steps, each aimed at the completion of a particular discovery task and accomplished by the application of a discovery method. The process iterates many times through the same domain, based on search in various hypotheses spaces. New knowledge is inferred from data, often with the use of domain knowledge.

A discovery system integrates various discovery methods, and can be used in interactive and iterative ways. Discovery systems can be compared by evaluating their autonomy and versatility. Autonomy measures to what extent a discovery system evaluates its decisions and produces new knowledge automatically, without external intervention. The degree of autonomy ranges from systems with low autonomy to assistant systems. Versatility measures the variety of application domains and steps of discovery processes that a discovery system supports.

To partially automate data exploration, discovery systems must perform tasks of content interpretation, which usually are still executed by an analyst who is an expert in the particular data domain. However, discovery systems should assist their users and not try to replace the user by automating tasks completely. To offer useful assistance and to take more tasks than existing systems do when supporting "manual" data exploration, they especially must be supplied with domain knowledge, be able to process imprecise instructions, and be able to adapt to a user's individual needs. Thus, discovery systems turn out to be a good example for the principles of an assisting computer.

We show how these assistance properties are realized in the system EXPLORA. First we give a short overview of EXPLORA and an application of exploring simulation results. We then refer to this application example to illustrate the domain knowledge used in EXPLORA and to exemplify the approaches to process imprecise instructions.

The Statistics Interpreter EXPLORA

The discovery system Explora (Hoschka & Klösgen, 1991; Klösgen, 1992, 1993, 1995a) supports the discovery of findings or unknown relations in data by searching for interesting instances of statistical patterns. A *pattern* is defined by a schema of a finding (Klösgen & Zytkow, 1995) and can also be seen as a generic statement with free variables. An *instance* of a pattern is a concrete statement in a high-level language that describes a hypothesis on data. Patterns must be *comprehensible,* that is, they should be understood directly by the analysts using a discovery system.

The high-level *pattern language* is a formalism to communicate new knowledge on a domain. The kind of statements constructed in such a language depends on the pattern type and varies from natural-language-type sentences like rules, to more abstract statements like trees, or even statements in a graphical language. An important component of a pattern language is the *concept language* used to build concepts.

The main dimensions of a *typology* for dependency patterns offered in EXPLORA are the number of populations compared in a pattern, the type of variables, deterministic or probabilistic verification methods, and the kind of language used to form concepts like subgroups, target groups, and populations of cases.

EXPLORA assesses *interestingness* of hypotheses and is computationally *efficient.* Problems arise because of the variety of patterns and the immense combinatorial possibilities of generating instances when studying relations between variables in subsets of data. The user must be saved from getting overwhelmed with a deluge of findings.

To restrict search with respect to analysis tasks, the user can *focus* each discovery run emerging during an interactive and iterative exploration process. Some basic organization principles of search can further limit the search effort. One principle is to organize search hierarchically and to evaluate first the statistical or information theoretic evidence of the general hypotheses.

Then more special hypotheses can be eliminated from further search, if a more general hypothesis has already been verified.

But this approach alone has some drawbacks and even in moderately sized data does not prevent large sets of findings. Therefore, in a second evaluation phase, further aspects of interestingness are assessed. A *refinement strategy* selects the most interesting of the statistically significant statements.

A second problem for discovery systems is efficiency. Each hypothesis evaluation requires many data accesses. EXPLORA uses special strategies that reduce data access and speed up computation.

Figure 3.5 illustrates the approach: EXPLORA constructs hierarchical spaces of hypotheses, organizes and controls the search for interesting instances in these spaces, verifies and evaluates the instances in data, and supports the presentation and management of the discovered findings. To process very large hypotheses spaces, several search strategies are available within a general search approach. A systematic but not exhaustive search cuts away whole subspaces, without skipping important hypotheses.

Findings are discovered by searching in spaces of hypotheses for instances of selected patterns. These instances must be interesting enough, according to some criteria.

Interestingness has several facets: *Evidence* indicates the significance of a finding measured by a statistical criterion. *Redundancy* amounts to the similarity of a finding with respect to other findings and measures to what degree

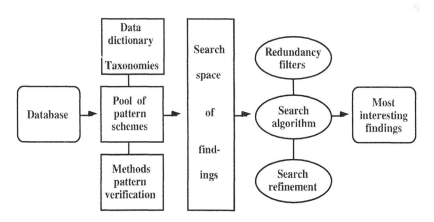

FIG. 3.5 Functional model of EXPLORA.

a finding follows from another one. *Usefulness* relates a finding to the goals of the user within the task under consideration. *Novelty* includes the deviation from prior knowledge of the user or system. *Simplicity* refers to the syntactical complexity of the presentation of a finding, and *generality* is determined by the fraction of the population a finding refers to.

Most of these facets are used for evaluation in EXPLORA. However, it is hardly practicable to assess the user's prior knowledge for measuring novelty. Moreover, the user specifies the analyses tasks simply by choosing the proper database, segmenting subsets, selecting certain groups of variables and records, and fixing pattern types and goals. Figure 3.6 shows a focus window of an EXPLORA application discussed in the following sections.

A prototype was implemented for the Apple Macintosh and used in a lot of practical applications (market research, medicine, election research, natural hazards, political planning, etc.), proving the potentiality of discovery systems. The implementation relies on Common LISP and CLOS (Common LISP Object System).

The implementation core of EXPLORA centers in the search algorithm and the generic specification of object spaces, patterns, operations, and hypotheses spaces. In case of database discovery, operations generate derived

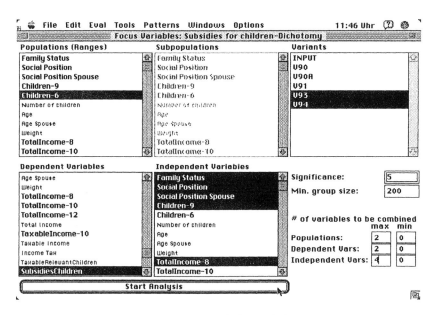

FIG. 3.6 Example of a focus window.

object classes like concept spaces defined by conjunctions of selectors in an attribute-value language or spaces of relations defined in a first-order language. For further discovery application fields (e.g., chemical or biological structures), other operations can be implemented.

A discovery system has to test thousands of hypotheses and process huge amounts of data within one discovery run. To overcome problems of resource shortages like computation time and main memory, the operations and representations of the derived object classes were implemented in a very efficient way using some special techniques. Therefore, EXPLORA 1.1[4] can be used in dialogue even for large data sets.

The central approach of EXPLORA is determined by a method for graph searching and by methods to construct a search graph. The search method can handle various pattern specifications (statement schemes) to identify within a graph those instances of a pattern (nodes representing hypotheses) that can be verified in the data. The graph to be searched for discovery of findings is constructed as a product set of partially ordered sets. These sets represent structures that are relevant for the application domain.

The Statistics Interpreter was also integrated into the environment of the assisting computer. A graphical user interface was implemented using the Generic Interactive Application (chapter 5, section 5.1), and the linkage to the knowledge-based Graphics Designer (section 3.3) was realized, transforming the identified findings of EXPLORA into appropriate graphical presentations. The knowledge base of the graphics designer contains rules for the determination of a valid diagram class and the construction of a diagram instance. These rules exploit some meta-data, that is, a description of structures and contents of the values belonging to a finding to be presented in graphical form.

Discovery and Simulation

The main discovery fields are knowledge discovery in databases (KDD) and automated scientific discovery (ASD). KDD centres around discovery in given databases. ASD deals with scientific comprehension by performing and analyzing experiments.

---

[4] To get EXPLORA 1.1, connect to *ftp.gmd.de* and transfer *Explora.sit.hqx* from the directory *gmd/explora*. *README* informs about installation. The "end user" version is available for applications on medium sized databases (up to 100,000 records). It is distributed as a stand alone program in object code.

ASD may feedback to data generation to improve the quality and scope of discovered knowledge. Further experiments in a laboratory generate new data. In many areas, experiments with a system to be investigated are impossible, but computerized models of the system are available. Data about the system are generated by running model simulations. Therefore, discovery in simulation results is endowed with the main characteristics of both KDD and ASD, that is, discovery in large databases and experimentation.

A model user controls the experimental conditions and defines a simulation *variant* by setting values for input variables and parameters. Two main questions are answered by experiments: What values do the output variables take, if a variant holds given values? How must the values of a variant be selected to achieve a desired output? A first analysis task refers to studying the results of a single variant. Factors are given and their influence on output variables are analyzed. For example, the consequences of a possible decision are calculated. Then discovery methods induce relations between input and output variables for the given variant, summarizing the various detailed relations of the model.

The second task is the comparison of two or more variants. Differences between output states belonging to the alternative factor combinations must be derived. Predictions under alternative conditions are made, or selections between already known decision alternatives are supported. Then discovery methods can derive significant differences between the variants for an output variable in subdomains of input variables.

Next, a whole space of variants is analyzed to derive general relations between factors and output variables. These relations clarify the process represented by the model. They are also used to plan decision alternatives. A fourth task deals with optimization or goal-state experiments. Combinations of factors are discovered that produce an optimal output. Often conditions for an output state are given and a variant is found achieving this state.

The main application fields of simulation are physical-technical and socioeconomic systems. Socioeconomic models are used in government agencies to plan legislation for taxation, transfer subsidies, health care, social security, and so on (Orcutt et al., 1986). The Model Base System, MBS (Klösgen & Schwarz, 1983), is widely used in German government to run macro-, cohort-, and micro models (Klösgen & Quinke, 1985).

Especially for microanalytic models, the volume of data is so large that a "manual" analysis of results is necessarily very restrained, and discovery approaches can offer a more complete and systematic analysis approach. The "manual" analysis is supported in MBS by standard queries with reports and some analyses (e.g., cross-tabulations).

The micromodels used for this EXPLORA application represent various tax and transfer laws (income tax, subsidies for households with children, subsidies for students) and operate on large datasets, such as a sample of about 50,000 German households. In a single simulation experiment, the model representing the law is applied for a given variant to each household. A variant includes a value for each of the parameters of the law. Output variables are derived by the model calculating output values for each household using input values and parameters. The main questions that are studied with these models are the determination of the total cost of a variant and the winners and losers of a planned legislative measure.

Domain Knowledge in EXPLORA

Discovery systems use data and domain knowledge to discover new knowledge about a domain. EXPLORA operates on four types of domain knowledge:

* Data dictionary knowledge
* Taxonomies
* Global statistical characteristics
* Specifications for interestingness.

Data dictionary knowledge includes labels for variables and their values, the types of variables (nominal, ordinal, continuous), the domain of variables, and "missing data" and "not applicable" specifications. EXPLORA relies on an internal data approach based on an inverted list data organization (Klösgen, 1995a). Initiated by an MBS-import menu option, the data dictionary knowledge belonging to a simulation model is generated, and microdata are converted into the EXPLORA data structures.

EXPLORA manages and accesses data and data dictionary knowledge within a hierarchy of applications, segments, and variables. Connected to simulation, an EXPLORA application refers to a simulation model and a segment to a variant. An input segment stores the input variables of a model that are variant-independent to avoid data redundancy in EXPLORA. Variables can be imported separately and further variants can be added.

Taxonomies are defined by an EXPLORA user and are hierarchies defined on the domains of variables. Often, they are introduced for discrete variables to restrict all possible internal disjunctions of values to meaningful disjunctions of a variable, and for ordinal or continuous variables to fix the intervals that are relevant within the application. For example, for the disjunction of the values "Hamburg," "Bremen," "Schleswig-Holstein," and "Niedersachsen" of the discrete variable "State," a node "Western-Coast-States" is defined. On the next hierarchical level, "Western-Coast-States" and "Eastern-Coast-States" are united in a further node.

A hierarchy of income-intervals can be introduced for instance, to represent the relevant intervals within the taxation domain. Therefore, taxonomies usually represent the preknowledge of the user about relevant categorizations within the domain. But additionally, a discovery system can cluster values of variables based on the results of discoveries. In this way, the system includes discovered taxonomies and concept definitions in its knowledge base and uses these concepts for further discoveries. EXPLORA operates on the hierarchies of taxonomies, especially during a first brute force process, typically performing a general to specific search.

Various *global statistical characteristics* can be used to support a discovery process. Methods to extract dependency patterns from data can, for instance, rely on dependency networks that describe the causal structure of variables (e.g., Spirtes et al., 1993). This and a lot of other statistical knowledge about data are generated as needed by the discovery system. For example, to capture such dependency information, EXPLORA evaluates the influence of independent variables on a dependent variable.

dichotomous relation	influence	continuous relation	influence
TaxableIncome	3534.2	Number of children	3320.8
Number of children	2509.1	TotalIncome	213.1
SocialPositionSpouse	1987.4	SocialPositionSpouse	89.3
FamilyStatus	1750.1	FamilyStatus	0.0
TotalIncome	0.0	TaxableIncome	0.0
.....		.....	

FIG. 3.7 Influence of independent variables on variable "subsidies."

Figure 3.7 shows evaluations for the dependent variable "subsidies for children" and some independent variables within the application "Micro Model: Financial support for families with children." This table analyzes the dependent variable as a dichotomous (support: yes or no) and a continuous variable (how much support). The results show that the influence of independent variables on a dependent variable can differ for some patterns. The statistical background of measuring the influence is not discussed here.

These characteristics can be used in a discovery run of EXPLORA to restrict the search process by including only the independent variables with a strong positive evaluation. This and other statistical knowledge about the data can be derived by the discovery system or can be introduced by the user when focusing a discovery process. During discovery focusing, the momentary focal point of interest is directed to a data section.

The application area "discovery in simulation results" differs from others by a high amount of knowledge that an analyst holds already about the domain and the process that generates the data, because the analyst knows the model and the law. Hence, for these applications the user will typically introduce much of this knowledge directly when focusing a discovery run.

The next category of domain knowledge includes several *specifications related to interestingness of findings*. At first one has to specify for a domain what types of findings (patterns) are relevant and which criteria are important for measuring the significance of a finding (verification method of a pattern). EXPLORA offers a general store of patterns that is nearly universally usable for data exploration, but that can be extended, if necessary, by further domain-specific patterns or adapted to special requirements of the domain. The adaptation of EXPLORA to a domain usually has to be done only once, enabling a user for a sequence of following discovery processes to select patterns and their verification methods by entering analysis goals and subgoals (see later discussion).

A second group of specifications refers to all kind of *preferences* that may be relevant for a discovery process. For example, preferences between independent variables can be specified (prefer TaxableIncome rather than TotalIncome, Number-of-taxable-relevant-children rather than Number-of-children). When evaluating interestingness of verified findings, EXPLORA uses higher weights for preferred variables (Gebhardt, 1994).

Preferences can be specified by users to adapt the discovery processes to their needs. On the other side, preferences could also be set by the system after monitoring and analyzing the work of the user. The user emphasizes the findings that are interesting for him or her and the findings that were expected (then deviations from expected findings are interesting). The system could then apply discovery processes to generalize these indications. In this way, the system could learn from the user and behave adaptively. Adaptive behavior has not yet been implemented in the current EXPLORA version. Proposals to assess user indications on interesting findings are described by Gebhardt (1994).

EXPLORA Processes Imprecise Instructions

To direct a discovery process to a special application problem, the analyst selects goals and subgoals and focuses a data section. A choice between main problem types is determined by two goals: number of variants to be analyzed (1, 2, $k > 2$), and qualitative (dichotomy or other discretization) or quantitative dependency. According to the possible combinations of goal specifications, a pattern is selected by EXPLORA for a discovery process.

Subgoals like "low classification accuracy," "high homogeneity," "disjoint findings," and so forth are introduced into EXPLORA, to select between different statistical tests for each pattern (a test evaluates the significance of a hypothesis) and several search algorithms. Some subgoals relate to the homogeneity, size, number, and strength of the concepts to be discovered. Other subgoals refer to the accuracy of the findings by fixing, for example, the granularity of the intervals built for continuous variables or other aspects of the language used to construct concepts, and the search strategy (exhaustive or stepwise). The degree of overlapping (or the disjointness,) and the focus on separation (e.g., minimal classification error) or key structure are fixed by further subgoals.

As two examples for these subgoal interpretations, we discuss now the rule pattern (goals: one population, dichotomous problem) and subgoals related to classification accuracy and disjointness of findings. The methods transforming imprecise instructions of the user into parameters and algorithms of the system are very simple in this case, relying on decision tables and simple rules. More complex transformations were introduced in chapter 2, section 2.3.

Typically for imprecise instructions in EXPLORA, the user specifies subgoals qualitatively (e.g., low, medium, high classification accuracy) and the system transforms this qualitative direction into a quantitative numerical parameter (e.g., an exponent in a formula). By tuning a slider (e.g., "still higher classification accuracy"), the user can modify a previous specification if not yet satisfied with the discovery results presented by the system. Of course, the user can also set a special value of the parameter if he or she knows about its theoretical foundation. Figure 3.8 shows the results of a discovery run for the two subgoals "high classification accuracy" and "low classification accuracy."

```
Micromodel:    Subsidies for children
One variant: V94
Problem:       Who gets subsidies?
Pattern:       Probabilistic rules
Households:    West Germany, with children,
               45% of the households get subsidies.

Subgoals:      High classification accuracy, Disjointness,
               Exhaustive search
Households with subsidies:
     95% of TaxableIncome < 30000
     94% of Married, Children 2, TaxableIncome 30000 -
            40000
     99% of Children 3, TaxableIncome 30000 -  62000
    100% of Children 4, TaxableIncome 30000 -  84000
    100% of Children 5, TaxableIncome 30000 - 106000

Subgoals:      Low classification accuracy, Disjointness,
               Exhaustive search
Households with subsidies:
     82% of TaxableIncome < 40000
     99% of Children 3,   TaxableIncome 40000 - 62000
     99% of Children > 3, TaxableIncome 40000 - 96000
```

FIG. 3.8 Rule sets for different classification accuracies.

To limit the amount of results, we refer in these examples to a comparatively simple law regulating the financial support of households with children. For the variant V94 of the law, 20 rules describe exactly in this simulation model who gets subsidies. Therefore, the differences between the subgoals "low" and "high" accuracy are not as distinct as in more complicated rule-sets holding, for instance, hundreds of rules.

The statistical evaluation of a rule is done in a two dimensional p-q space, with p as the probability of the rule, that is, $p = P(RHS \mid LHS)$, and q as the relative frequency of the conditional part (concept), that is, $q = P(LHS)$. Let $p_0 = P(RHS)$ be the relative frequency of the conclusion, which is fixed because we regard all rules with a given right-hand side (RHS). The admissible section in that p-q space (due to constraints) and some isolines of the following evaluation functions are shown in Fig. 3.9 with the following evaluation functions associated to the three qualitative default values in EXPLORA:

(1) $E1 = q\,(p - p_0)$ (low accuracy)
(2) $E2 = (q\,/\,(1\text{-}q))\,(p - p_0)^2$ (medium accuracy)
(3) $E3 = \sqrt{q}\,(p - p_0)$ (high accuracy)

Evaluation (3) is equivalent to the binomial test. One can show (see Klösgen, 1995b) that (2) is equivalent to the criteria of CART, Gini index, chi-square test for 2X2 contingency tables, and to the INFERULE (Uthurusamy et al., 1991) criterion for two classes (all these criteria are equivalent); (1) is equivalent to a criterion that was introduced by Piatetsky-Shapiro (1991) as the simplest criterion satisfying some basic principles. Similar results are

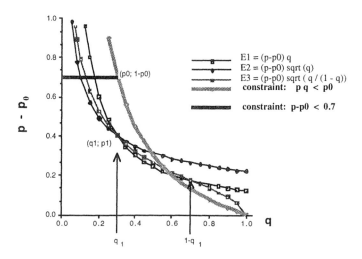

Example: $q_1 = 0.3$, $p_1 = 0.7$, $p_0 = 0.3$

FIG. 3.9 Isolines for evaluation functions and constraints in evaluation space.

obtained for the cases of general discrete and continuous dependency, where variance reduction or homogeneity is used as subgoal.

Equivalence is defined related to allowed transformations on the evaluation functions. The equivalence given earlier refers to possible multiplications with constant factors (constant with respect to p and q). This kind of equivalence is important for the refinement algorithm of EXPLORA, which is invariant to such multiplications.

Perhaps the user is not satisfied with the default specifications of this subgoal used in Fig. 3.8 and requests a still higher accuracy. The system interprets the actual parameter and the qualitative direction of the user and further decreases the exponent in (3). Figure 3.10 shows the results that belong to an exponent 0.25 (fourth root of q).

Another equivalence can be defined with respect to maximum preserving transformations. In case of disjoint rules, maximum preserving transformations of evaluations are allowed, because in this mode the simplest strategy is to iteratively select the rule with the maximal evaluation, discard all overlapping rules, and select the next maximal rule. This strategy is applied in a similar way by the CN2 algorithm.

```
Subgoals:      High classification accuracy
               (increased: E=0.25),
               Disjointness, Exhaustive search
Households with subsidies:
100% of TaxableIncome < 18000
 90% of Married, Children 1, TaxableIncome 18000 - 30000
 97% of Married, Children 2, TaxableIncome 18000 - 40000
 95% of Single,  Children 2, TaxableIncome 18000 - 30000
100% of Married, Children 3, TaxableIncome 18000 - 62000
100% of Single,  Children 3, TaxableIncome 18000 - 52000
100% of Children 4, TaxableIncome 18000 -   84000
100% of Children 5, TaxableIncome 18000 - 106000
```

FIG. 3.10 Classification accuracy further increased.

The practical relevance of adjusting this subgoal depends on the complexity of the law. More complex laws hold a large number of regulations. The probabilistic rule pattern can then be used to generalize these regulations gradually, by adjusting classification accuracy, and to uncover the dominant structure of the dependency.

A second example of qualitative subgoals refers to the degree of overlappings of findings. In the preceding examples, no overlappings (disjointness) was specified as subgoal. This corresponds to an exponent $k = 0$ in an algorithm[5] suppressing overlappings during search refinement. This suppressing procedure is studied in detail by Gebhardt (1991). If the user selects "low overlapping" as subgoal, this qualitative setting is transformed to a parameter $k = 0.25$ and the results of Fig. 3.11 are discovered.

The quantitative parameter determined by EXPLORA for a qualitative subgoal specified by an user must hold some properties to guarantee natural and unmanipulatory results. The most important property satisfied by these methods is continuity. That is, small changes in the values of the parameter may not lead to large differences in the presented results.

```
Subgoals:    High classification accuracy, Low overlap-
             ping, Exhaustive search
Households with subsidies:
   100% of TaxableIncome < 18000
    95% of TaxableIncome < 30000
    98% of Children 2, TaxableIncome < 40000
   100% of Children 3, TaxableIncome 18000 -  62000
   100% of Children 4, TaxableIncome 18000 -  84000
   100% of Children 5, TaxableIncome 18000 - 106000
```

FIG. 3.11 Rule set for a different level of overlapping.

Conclusion

To efficiently support an analyst in database discovery, a discovery system must be realized as an assistant system. The assistance paradigm is important because discovery is an interactive process that depends on the goals of the analyst and emerges iteratively. Because this process cannot be specified in advance, a discovery system cannot replace an analyst by automating discovery completely. On the other hand, "manual" exploration of large datasets, that is, testing single hypotheses specified individually by the user, does not

---

[5] EXPLORA uses the following *affinity function* for the extensions of two subgroups (V is evidence): Definition: $S(M,N) = (|M \cap N| / |M|)^k$. A finding F_j suppresses a finding F_k, if $V(F_k) < S(F_k, F_j) V(F_j)$.

meet the requirements of "data mining." Discovery systems must assist the analyst in generating, searching, and evaluating spaces of hypotheses.

EXPLORA operates on data and domain knowledge to discover new knowledge about a domain. Data dictionary knowledge, taxonomies, global statistical characteristics, and interestingness specifications constitute the domain competence within EXPLORA. Some of this knowledge can also be generated by discovery. Future generations of systems will include discovered knowledge in their domain knowledge to a still higher extent and will use these findings for further discovery processes. EXPLORA processes imprecise instructions by transforming qualitative goals specified by the user into numerical parameters of algorithms. Such goal specifications are necessary to adapt discovery processes to the needs of the user and the domain. Future, more advanced discovery systems will incorporate more learning and adaptive behavior. Discovery methods should be used also to learn from the user by monitoring and analyzing his reactions on the discovered and presented findings to assess the novelty facet of interestingness.

3.3 GRAPHICS DESIGNER

Klaus Kansy, Dieter Bolz, Rüdiger Kolb, Günther Schmitgen

The diffusion of word and graphics processors over the past decade has improved the technical quality of documents, but not their design. Although the very perfection of the printed page raises readers' expectations, it also draws attention to any deficiencies. Tools like spelling checkers and automatic hyphenation help control low-level aspects of the text; and style guides can format a document according to a given style more or less automatically. For the graphics side, something equivalent is still missing. Much time still has to be spent with the boring task of arranging, aligning, and resizing graphics elements individually and relative to each other.

The visualization of numbers and related data is a basic task of statistics packages. The statistics interpreter (see section 3.2) is a good example for illustrating the need for visualisation tools. Spreadsheets and statistics packages commonly offer tools for generating charts and often provide tutorials in their use. Yet one study found inappropriate graphics in 24 out of 50 U.S. corporate annual reports, even though such reports are designed carefully by professionals (Johnson et al., 1980). Users are left alone with a wealth of tools and little guidance on when to select which presentation technique or stylistic means for specific data in a specific situation.

Education of authors is not the solution—at least not the ideal solution. Authors are experts for their application and not for graphics design. They should be able to focus on composing content without having to worry much about design questions.

This is where the so-called GRAPHICS DESIGNER comes in. The GRAPHICS DESIGNER assists an author in various aspects of composing graphics. As an ever-present servant, it releases the author from trivial tasks like precise positioning and aligning. For more complex design problems, the GRAPHICS DESIGNER cooperates with the user to find a solution; it points to possible design problems and gives hints on how to solve them. The GRAPHICS DESIGNER controls the selection of graphical and stylistic elements according to the given data and intentions of the user.

The design knowledge of the GRAPHICS DESIGNER will always be limited. With this limitation in mind, we looked at the way humans try to cope with a new design problem. A common approach is to browse through a set of good samples (one's own or from textbooks) to get some ideas and a starting point for the current task. The GRAPHICS DESIGNER mimics this approach and backs up its built-in help facilities with databases of examples of good practice. A user can formulate a query with rough sketches to search the graphical database for similar forms. This search tool is useful and meaningful for maintaining and accessing huge picture databases.

The State of the Art in Design Tools

To build a GRAPHICS DESIGNER one has to collect knowledge about graphical design and formalize it in a way that a computer can work with it in a meaningful way. As source for such knowledge, Bertin (1983) is a standard reference for all people working in the area. Bertin looked at the different graphical means as a graphics vocabulary comparable to other notations. He identified visual variables (e.g., coordinate values, size, shape, color), classified them, and described their expressiveness. Tufte (1983) introduced the terms *graphical excellence* to describe global design aspects, *graphical integrity* to demand a correct representation of data, and *data/ink ratio* to distinguish required graphical elements from redundant decoration. Zelazny (1986) became an example of a tutoring book for business graphics that gives concrete rules how to conclude the relation and chart type from data. These rules can easily be transformed into rules for computers.

To store such information in a computer, special graphical languages are needed that allow the representation of the required knowledge. Mackinlay (1986) is a standard reference for the basics of a formal graphics language to represent relational information. A successful implementation for a specific application area (instruction manual for a radio) has been done by Feiner (1988), who used a rule-based system for generating appropriate visualizations to draw the focus of attention to the relevant parts of the radio described in the manual. Similar work was done within the WIP project of the DFKI, where layout rules for multimodal documents were implemented (Maaß, 1992). Chappel and Wilson (1991) described recent work on a multimodal user interface where the graphical representation of application data is generated by a rule system. The rule formalism is based on Prolog and allows neither imprecise rules nor rule retraction.

Geometrical aspects in graphics are expressed by analogical (coordinates) rather than propositional values. Coordinates are well suited for the generation of a display but are not the way humans describe graphics. Reasoning about analogical values is difficult; the adaptation of Allen's temporal logic for space (Allen, 1983) is an often-cited approach, which, however, allows only for a very limited reasoning about relative positioning of objects. Constraints as introduced by Borning (1979) in his ThingLab system are another way to express high level concepts for analogical values. Borning's ideas were further developed by his group for incremental constraint solving (Freeman-Benson et al., 1990) and hierarchical constraints (Borning et al., 1987).

Currently, systems appear on the market that lead into the direction of more user-friendly and intelligent graphics systems and behave in some aspects like a design assistant. The graphics capabilities of pen-based computers like Apple's Newton include astonishingly effective pattern and character recognition software to interpret rough sketches made by pen. New graphics editors like the drafting assistant from Ashlar Vellum (1992) allow the specification of connection points, attraction points, and so on, which simplifies the generation and manipulation of graphics for a user who can properly handle such concepts.

Our Approach: The GRAPHICS DESIGNER

The GRAPHICS DESIGNER (Kansy, 1990, 1991) is a software system that supports the design and the beautification of business graphics at a high conceptual level. Knowledge about the content is used for choosing the adequate presentation form. Built-in graphics design knowledge allows the user to be relieved from boring routine tasks and gives guidance in the use of graphics presentation means.

Existing graphics can be retrieved from pictorial databases by using graphical search criteria, which may be given by a sketch. Salient parts of a graphics will be recognized for beautification or for a content-based graphical retrieval.

Graphics design allows for a great variety of valid solutions. Rather than enforcing a certain design policy or asking the user for a unique and complete specification of the desired result, the GRAPHICS DESIGNER uses its design knowledge in conjunction with the given data and user-specified intentions to eliminate invalid and to offer recommended solutions. This behavior reflects a view of the design task as a cooperative and iterative process where an autonomous user brings in what he or she knows and wants to provide and the assistant design system tries to complete the task as far as is possible and sensible. This allows the design system to perform routine tasks automatically, to control the consistency of the result, and to cooperate with the user on tricky problems.

In graphical design, vagueness of various kinds is inherently present. The user may input a rough sketch with imprecise positions and sizes that have to be aligned. The system has to recognize which values are independent and can be changed immediately and which values repeat in the graphics and can be changed only in a coherent way. Multiple solutions are common in design, and it does not make sense to enforce a specific one by the system. The user may specify only part of the parameters, which leaves it to the system to complete the definition by meaningful default values. The user may specify too many conditions, which cannot be fulfilled within one design, such that the user has to withdraw some requirements to make a solution possible. The adequate treatment of these kinds of vagueness and inaccuracy impose strong demands on the methodological basis.

As a prerequisite of the implementation, a flexible object-oriented graphics system was developed that gives full access to all graphics attributes considered. This allows artificial intelligence tools to perform reasoning about features present in the graphics and to modify them.

The work was concentrated on selected areas to allow for a sufficient deep study and prove of concepts by prototype implementations. Four prototypes have been developed as components of the GRAPHICS DESIGNER:

1. The 1BEAUTIFIER transforms sketches into finished graphics and improves the graphical design quality; it can be regarded as a kind of graphical spelling checker.

2. The CHART GENERATOR generates charts from given data and intentions. It can work with incomplete descriptions and serves as a consultant for chart selection.

3. The GRAPHICAL SEARCH component allows the retrieval of graphics according to graphical search criteria. The criterion can be a sketch that indicates the desired features of the intended graphics.

4. The GESTALT RECOGNIZER recognizes salient features (so-called "Gestalten") of graphics. This component is described in section 3.4.

In the following chapters, we describe the graphics language EPICT, serving as a foundation of the GRAPHICS DESIGNER, and the components BEAUTIFIER, CHART GENERATOR, and GRAPHICAL SEARCH.

EPICT: GRAPHICS DESIGNER'S Language for Spatial Concepts

Classical graphics systems describe graphics by coordinate values for geometrical aspects and numbers for attributes like color and pattern. This is sufficient for a display-oriented graphics system because it can readily generate the display from the given data.

If a human looks at graphics, other descriptions are important beyond mere numbers. Spatial relations between objects (e.g., does_contain, is_horizontal, are_intersecting, are_aligned, left_of, etc.), numbers of similar objects, groupings of objects, characteristic shapes, and so forth play prominent roles. To analyze graphics for such features and to reason about such notions with a reasoning system, a graphical language is required that is able to express such concepts. A further problem is to maintain the relationship between the analogical(coordinate-oriented) and propositional notions as far as possible.

These ideas led to the definition and implementation of a new object-oriented graphical language, EPICT (Rome et al., 1990). EPICT is built up in layers where the lowest layer is a pure analogical level and covers the functionality of a typical two-dimensional graphics editor (e.g., primitives like point, line, arrow, text, polygon, conic, rectangle, bitmap, spline). It contains geometric objects in the mathematical sense (e.g., with line width 0) and presentable objects where the geometry is inherited from the geometric objects and display attributes are added. The next layer contains auxiliary functions for single objects like bounding_box, center, and attributes like is_horizontal, is_square. The third layer contains interobject relations like are_intersecting, are_parallel.

Building on these layers, more complex concepts can be added that give access to all interesting graphical attributes and relations within graphics. One example is the so-called situation language (Bolz, 1990), which allows for a declarative description of elements in graphics that are searched for retrieval or correction purposes. An important feature of the situation language is its capability to describe "vague" situations, that is, to allow ranges for parameters. By this feature, attributes like approximate_equal, nearly_horizontal can be modeled or a variable number of items can be looked for.

Beautifying Draft Graphics Through Automatic Design Analysis

The BEAUTIFIER (Bolz, 1993b) analyzes draft graphics according to aesthetical and design criteria and corrects inconsistencies found. It concentrates on geometrical (e.g., positions, sizes, alignments) and stylistic elements (e.g., use of fonts and patterns).

Modern graphic editors offer a lot of functionality in a comfortable way. However, there are only few means available to assist the user in trivial beautification operations. Grids are a common support tool for the regular positioning of objects. However, the limitations of a grid are obvious; for example, they do not allow the precise positioning of objects at the corners of an equilateral triangle. Bier developed a technique called "snap dragging" (Bier, 1989), in which a sort of gravity mapping makes the cursor snap to particular points, lines, or surfaces, depending on the context. The drafting assistant of Ashlar Vellum (1992) anticipates during the construction process which snap points and snap lines could be useful and offers them to the user. Pavlidis and van Wyk (1985) described an automatic beautifier that analyzes graphics and generates constraints that are to be satisfied to achieve beautified graphics. Their concern was the technical problem of beautification and not the user interface, which is essential for a task that usually cannot be fully automated.

The BEAUTIFIER developed as part of the GRAPHICS DESIGNER does not aim at an automatic service (Bolz, 1993a) because a beautifier—in our opinion—never can be perfect. Even if a correction is useful, it would be surprising for a user if a correction happened during the design process. Following the assisting metaphor, the user may use the initiative to ask the BEAUTIFIER for an analysis of the drawing. The BEAUTIFIER then proposes corrections, which the user may accept or reject. The functionality of the BEAUTIFIER is based on distinct rule sets, which the user can select individually to direct the beautification process.

The beautification process is performed in a cycle of analysis, critique, and modification (see Fig. 3.12) iterated under user control.

For the analysis of graphics, error situations in graphics are defined using a flexible declarative situation language (Bolz, 1990) based on EPICT. Typical error situations are predefined; further situations can be added by a programmer using the constructs of the LISP-based situation language. The expressiveness of the language has proven sufficient for the given purpose.

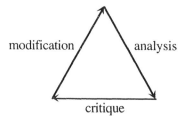

FIG. 3.12 Beautification as a cycle of analysis, critique, and modification.

All situations in the graphics corresponding to the selected set of error situations are found automatically and then transferred to the critique module. Errors may relate to other errors, such that the correction of one error influences other errors. The critique module has the task to carefully select the errors to be corrected, to avoid undesired cycles.

The correction of situations could be specified by a system of constraints describing the desired target states. However, a constraint solver will run into difficulties in dealing with the kind and number of constraints typical for graphics. A further difficulty is that the solvability of the constraint system cannot be guaranteed.

Therefore, we have chosen a different approach in our implementation. The description of each situation is completed by one or more operators that correct the inconsistency. For each operator the side effects are noted: the parameters it affects and the parameters it depends on, such that a subsequent change could destroy the effect of the operator. A set of operators can then be chosen incrementally where the single operators do not interfere with each other. The selection is done by search over all possible operators, while carefully keeping track of all side effects. The search can be interrupted at any time and the found operators can be applied. These operators will usually improve the quality considerably. Then the whole process can be repeated. Precautions have been taken to avoid cycles.

This approach has several advantages with respect to assisting properties: The system performs operations where each step is meaningful for the user, rather than performing a lengthy computation and confronting the user with the final result only. By performing each operator, the graphics will change in an understandable way and normally improve in quality. Therefore, the process can be interrupted at any time, delivering an intermediate but improved result. It will solve all trivial inconsistencies immediately and iterate until the graphics is corrected or just the serious problems remain, which are

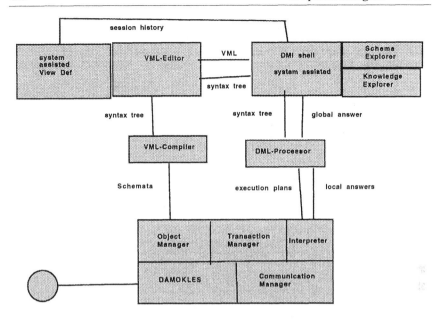

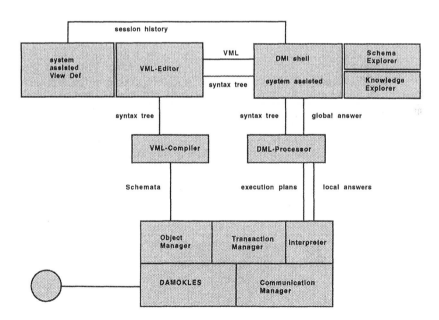

FIG. 3.13 The effects of the BEAUTIFIER on a sample drawing.

brought back to the user for advice and help according to the assistant paradigm.

Figure 3.13 shows a nontrivial example of the effects of the BEAUTIFIER.

Generating Appropriate Charts from Vague Hints

The CHART GENERATOR (Kolb, 1993) is a component of the GRAPHICS DESIGNER that accepts data and rough hints about the desired display from the user and generates automatically—with possible interventions by the user—an appropriate and correct chart.

Charts are in widespread use; however, the usage is often based on arbitrary decisions rather than governed by a deep understanding of the semantics and aesthetics of charts. A specific type of chart can only be used for specific types of comparisons; for example, time series should be represented only by horizontally oriented charts (line charts or vertical bars) (Zelazny, 1986). Charts are complex objects where graphical parts are associated or inter-leaved with descriptive text; charts obey nontrivial rules and are difficult to optimize with the common general-purpose chart generators. Typical errors include labels that are difficult to read or, worse, the choice of an inappropri-ate chart type just for the purpose of easy labeling. An assistant that controls the graphic and semantic integrity of a chart and warns the user of critical or wrong decisions would be of great benefit for the clerical worker charged with the generation of the chart.

The CHART GENERATOR tries to fulfill these requirements. It translates the description of data and intentions into a formal description of the chart from which the chart can be constructed. This task is performed with the help of a rule base containing design rules derived from standard text books on graphics design (e.g., Bertin, 1983; Tufte, 1983; Zelazny, 1986) and by interviewing a human expert.

The data and intentions for our prototype system are delivered by the statis-tics interpreter EXPLORA (see section 3.2) which analyzes real statistical data and extracts interesting findings. An example output is the subdivision of the clients of a bank according to educational level, and the finding that clients with higher educational level make heavier use of a bank depot. To produce charts from such data, application specific rules and a taxonomy of the application terminology have also been included to connect the applica-tion with the design world (Kolb, 1992).

The generation of a chart is implemented as a multistage reasoning process (Fig. 3.14) where, in the first phase, the type of comparison is derived, and then the possible chart types are selected and a prototype chart is generated. Finally, all aesthetical refinements are performed that depend on the concrete extent of the graphics and textual components.

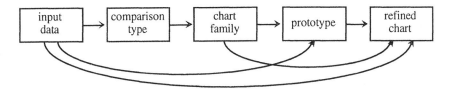

FIG. 3.14 Chart generation: from input data to finished chart.

The CHART GENERATOR starts with the data and all available information about the kind of data and the intention of the desired chart. This information may include type of data (e.g., nominal, ordinal, quantitative, percentage), meaning of data (e.g., time, currency, weight), or intentions conveyed by the chart (e.g., "emphasize maximum value"). Such information is easily provided by the user, who will specify only facts about the data and the purpose of the chart and will remain in his or her area of expertise. Furthermore, there are now systems available that automatically generate data with such characterizations. The statistics interpreter EXPLORA described in section 3.2 is an example of such a system.

In the first phase, the type of comparison is determined. According to Zelazny (1986) there are five types of comparison: structure (decomposition of an aggregate), rank (ordering of data), time (change of data over time), frequency (distribution of data), and correlation (dependencies among data). Either the type of comparison can be deduced from the kind of data (time data indicates a time series) or the kind of intention ("show maximum" indicates rank), or the possible types of comparison can be restricted.

Each type of comparison can be associated with one or more chart families. The chart families are pies, horizontal bars, vertical bars, line diagram, and point diagram. Zelazny gave a mapping that has been implemented. As other textbooks gave slightly different choices, this mapping is not regarded as an ultimate rule but handled as a recommendation with high priority.

Within a chart family like bar charts there are several different graphical realizations, for example, two- or three-dimensional, with label on, above, or below the bar, and so forth. The third phase tries to narrow down the possible choices, such as by looking at the size or number of data and the length of labels. Then the description is complete enough to allow drawing the chart.

In the last phase, those aspects are controlled and modified that depend on the graphical image of the chart. Examples include overlap of labels and positioning of title within the available free space.

The foregoing description of the working of the CHART GENERATOR shows that a standard rule-based system would not be the right choice for implementing the intended functionality. Standard rule-based systems require a complete, a priori definition of the task and operate on a well-defined set of valid rules to obtain the answer. An important feature of this CHART GENERATOR is that the user is not forced to fully specify the task. The user provides as much information as he or she wants to, such as the information immediately available, and the CHART GENERATOR completes the description by default knowledge. Naturally, the less information is given, the less precise and directed the result will be. At least the CHART GENERATOR will avoid presentations that contradict the given information. With no additional information at all, just the most common chart should be selected. This generous handling of partial specifications can be achieved by a default reasoning system that allows the distinction of strict rules and possible default rules, where the latter apply only in the absence of other information.

A further observation was that strict rules are not typical for graphics design—on the contrary, they are more or less the exceptions. Usually, design rules allow for different possible alternatives, the merits of which can be equal or ordered by priority. So we need a system that allows assigning priorities to rules.

Questions of taste are important in selecting a specific design; this suggests asking for a system where different design alternatives can comfortably be explored and compared. Furthermore, the process of design includes some preliminary decisions that may be withdrawn later on if they do not lead to convincing results.

The nonmonotonic reasoning system EXCEPT II (Junker, 1991; see chapter 2, section 2.3) was chosen as basis for the CHART GENERATOR. EXCEPT II distinguishes between *rules* for firm knowledge and *defaults* for preliminary knowledge and expresses preferences by assigning priorities to default rules. With this system it was possible to represent the chart design knowledge and to model the incremental generation of the desired result. When starting the reasoning system with the initial set of data, the result normally will not be fully determined. The CHART GENERATOR will automatically select defaults with highest priority to complete the formal description to a plausible solution. The charts are generated and presented to the user. The user can then look at alternative solution by asking for alternatives with lower priorities; can explicitly exclude certain default assumption that he or she does not like in this context; and can add at any stage further information to guide the process of selection.

It is this kind of behavior that characterizes the assisting role of the CHART GENERATOR. System and human operator cooperate as autonomous systems in a reasonable way each providing special capabilities. The user can freely express creativity by adding any rules and can explore the space of possible solutions. On the other side is a competent system that helps an autonomous user to cope with the task; it does the bookkeeping, fills up undefined slots by plausible defaults, and controls compliance to general design rules, preventing the user from exploring invalid solutions.

The metaphor of cooperation makes such an ambitious system possible because there is no need for a (practically impossible) complete modeling of the design domain. The system does reasoning as far as the implemented knowledge allows and leaves everything else to the human partner. A prototype implementation has verified the viability of this concept.

Searching for Model Graphics Using Rough Sketches

Graphical search is an important counterpart to the generative components of the GRAPHICS DESIGNER. This function complements the classical, keyword-oriented retrieval techniques for graphics by giving the user the possibility of specifying a query graphically by drawing a rough sketch of significant features of the desired graphics. A sketch is—by definition—rough and imprecise; the retrieval mechanism working on such a query has to cope with vagueness in position, size, number, and so forth of the graphical objects within the sketch. Even the type of the objects may be inexact: A rectangle can be one integral object, combined from four lines, or drawn by freehand input.

A graphical retrieval based on an inexact sample is also an appropriate means of finding relevant examples for a design task. The user sketches the major components of a figure, and the search component finds similar figures by searching a graphical data base with professionally designed samples (see Fig. 3.15). This facility complements the generation and beautification components in areas where no relevant knowledge could be acquired and formally represented.

Today, pictures in professional graphical databases are classified and assigned appropriate keywords by experts. Then classical text-oriented retrieval systems can be used for graphical search. Our approach aims toward a clerical user who has no time or interest to systematically classify a set of graphics and who asks for a kind of graphical "full text retrieval."

Existing solutions for graphical retrieval have only limited scope (see survey in Henne & Schmitgen, 1991). Fellner and Stögerer (1988) searched for specific graphics primitives and attributes in a standardized picture format (CGM), which typically is not the terms in which a user remembers a figure. Spatial relations of graphical objects seem more appropriate for humans to describe graphics. Such terms are studied in the context of spatial reasoning systems. Allen's (1983) temporal logic for describing all possibilities of temporal overlap has been extended to spatial orderings (Güsgen & Fidelak, 1990). Hernandez (1990) proposed a system for describing the direction and the distance from one object to another, which, for each object, gives an

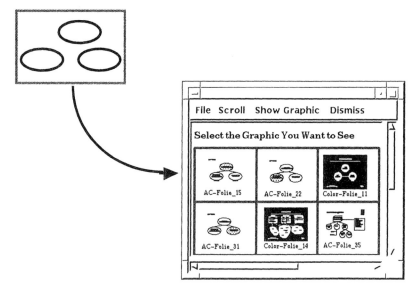

FIG. 3.15 Search by sketch.

abstract map that allows for limited reasoning about relative spatial positions of objects. Analogical methods retain the geometrical information in a reduced form for comparison purposes or algorithmic processing: mapping of graphics to a coarse grid (see chapter 2, section 2.3) gives a rough feature map that codes simplified geometrical information. Two-dimensional strings (Chang et al., 1987) are another way of extracting ordering information. However, the granularity can only be chosen arbitrarily, which limits the applicability of such approaches.

As we could not find one general and all-encompassing scheme, we chose to implement a hybrid system (Henne & Schmitgen, 1993) that combines features from several systems. It automatically extracts information graphical primitives with propositional (type of primitive, number of primitives of a specific type, etc.) and positional features (location within graphics, relative position, etc.) and their graphical attributes (color, line style, etc.). Text within graphics is used as valuable keywords. A very novel approach is the analysis of graphics for so-called Gestalt features (Rome, 1994) (see section 3.4). Gestalten represent characteristic structural elements in graphics that are of cognitive importance for the visual perception of humans.

All these features are extracted from each graphics in the data base and stored in an Associative Memory Model (ASM) (Henne, 1990) (see chapter 2, section 2.3). This is a time-consuming process that is done off-line. A query is specified via a graphical sketch. The set of features is extracted from the sketch and sent to the ASM, which determines exact or best matches within the stored graphical data base. If the number of hits is to big, the user has to narrow the query. Otherwise, the hits are presented in the form of miniaturized graphics for visual checking and selection by the user (see Fig. 3.15).

Conclusion

The GRAPHICS DESIGNER is a software system that supports the design, the beautification, and the search of business graphics at a high conceptual level. It was an enabling project that tested the viability of different approaches for supporting a clerical worker in designing graphics. The challenge of the project was the acquisition of graphics design knowledge in a form suitable for machine processing and the application of methods for reasoning under uncertainty to cope with an underspecified or contradictory problem description. The assistant metaphor required for flexible handling of the knowledge, awareness of the limitation of the system, and appropriate inclusion of the user in the generation process. It showed typical functionality to be expected from graphics systems in the future. "Intelligent" graphics systems appearing on the market and the advent of pen-based computers, like Apple's Newton, that deal with "vague" graphics input have shown that the problems tackled with the GRAPHICS DESIGNER are relevant for the development of future computer systems.

Success depended heavily on interdisciplinary work of experts in the area of computer graphics and artificial intelligence. Techniques of nonmonotonic reasoning and planning under uncertainty were applied to achieve the ambitious goals. The experiences showed limitations of current reasoning tools and influenced their further development.

As a side effect, a flexible object-oriented graphics system EPICT was developed, which gives reasoning tools full access to all graphics attributes under consideration. A prototype of the GRAPHICS DESIGNER has been implemented.

Acknowledgment

The GRAPHICS DESIGNER activity was started in 1988 and ended in 1993. It has been part of the joint project TASSO[6] (see chapter 2, section 2.3) where its methodological basis has been developed. The authors acknowledge the contributions made by all members of TASSO to the work described.

---

[6] The project TASSO was partially funded by the German Federal Ministry for Research and Technology under contract ITW8900A7.

3.4 GRAPHICAL SEARCH ASSISTANCE

Erich Rome

In this section we propose a method for the detection of salient nonlocal structures in presentation graphics. Nonlocal structures may consist of similar graphical objects—the constituents of a presentation graph—or of objects that are arranged with order. They may be perceived immediately, but they are not explicitly represented in the internal description of a graph. Information about such cognitive relevant structures may be used to guide the operations of the beautifier component of the Graphics Designer system and may serve as additional indices to the graphics database of the Graphics Designer's graphical search component.

The method used to detect nonlocal structures is the simulation of models of organizing phenomena of human visual perception. Nonlocal structures emerge as a result of grouping processes of visual perception. Adherents of the Gestalt school of psychology have investigated these processes and have found a number of principles that guide them. In the last decade, some promising models for the quantification of the rather qualitative Gestalt principles have been published. Among these are Treisman's Feature Map Model of visual perception and Palmer's Transformational Approach to visual perception. They provide a basis for the computer simulation of grouping processes and are used in our own MAX simulation.

MAX simulates the grouping of graphical objects according to the Gestalt grouping principles of proximity, similarity, and good continuation. Groups of primitive objects may in turn themselves be grouped in a hierarchical fashion. One important aspect of the form of a group is the arrangement of its constituent parts. Such arrangements are represented using polygons, with their vertices being the center points of the group's elements. Arrangement polygons are compared by a polygon matching algorithm that computes a similarity metric for polygonal shapes; this makes it possible to retrieve similarly arranged groups in the Graphics Designer's database of presentation graphics.

An Example

The analysis of presentation graphics, as performed by the graphical search component (Henne & Schmitgen, 1993, cf. chapter 5.) and the beautifier component of the Graphics Designer system (Bolz, 1993b, cf. section 3.3), lacks one important property: Both components have only limited capabilities for the detection of *nonlocal structures.* Nonlocal structures, such as four rectangles of equal dimensions and colors, which are equidistantly positioned on a horizontal line, may be perceived immediately, but they are not explicitly represented in the internal description of the graphics.

Furthermore, because such structures can be found in many presentation graphics and because they may be salient and of strong visual appeal, they seem to be of some cognitive relevance to a user of the Graphics Designer system (Kansy, 1991, cf. section 3.3). Thus, it seems desirable to introduce a method for the detection of nonlocal structures in order to make them available to other components of the Graphics Designer system. Enabling the system to deal with cognitive relevant nonlocal structures is expected to improve its assistance capabilities.

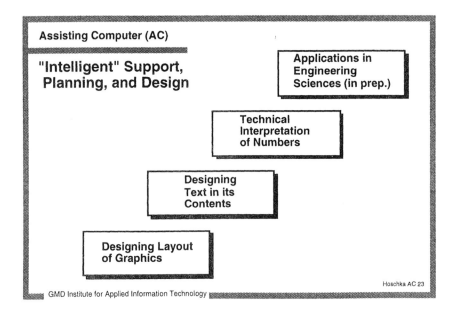

FIG. 3.16 A typical presentation graphic.

Figure 3.16 shows a typical presentation graph. The dominating global structure is the diagonal arrangement of shadowed text boxes. They form a group according to the Gestalt principles of similarity and of good continuation (described later). The constituent local elements of this group are represented as primitive rectangles and text elements in the Graphics Designer's picture representation language EPICT (Rome, 1991), but there is no explicit representation of the global group as a whole.

Arrangements like this one may serve as additional descriptions, that is to say, as indices to a graphic in the search component's database of presentation graphics. A method for the automatic detection of nonlocal structures may be smoothly integrated into the graphical search component as described in the following:

First, presentation graphics must be analyzed with our method when they are stored into the database. Each detected global group has to be stored as an additional index to the corresponding graph. When the search component receives a query in form of a search sketch, the sketch has to be analyzed in the same way. Any found group in the sketch is then compared with the group indices in the database. All found graphics with matching group indices are handled by the search component in its usual way: They are presented as scaled-down *thumbnails,* and if the searched graph is among the displayed ones, a user may retrieve it in its original size by clicking on the corresponding thumbnail.

The question then is: How can a program detect global arrangements automatically? The method proposed here to detect nonlocal structures is the simulation of models of organizing phenomena of human visual perception. Nonlocal structures emerge as a result of *grouping processes* of human visual perception. Exponents of the *Gestalt school of psychology* have investigated these processes and found a number of principles that guide grouping processes (see Ellis, 1974). Among these are the grouping of local elements to global wholes of different quality by *proximity, good continuation,* and *similarity* of form, color, brightness, and other features.

The remainder of this section is structured as follows: In the next subsection, the effects of the Gestalt principles just listed are demonstrated using some instructive examples. This is followed by a summary of the state of the art in application-oriented computer simulations of models of Gestalt grouping processes. Thereafter, a description of the MAX grouping and matching algorithms is given. Some test results on artificial patterns and real presentation graphics may give an impression of the performance of MAX. We conclude with a critical summary of the approach and with some hints for future directions of research.

On Gestalt Perception

Although the heyday of the Gestalt school of psychology of the 1920s and 1930s is long gone, the Gestalt principles of visual perception are still very popular and have exerted some influence on current research. Some authors have termed this phenomenon "Neo-Gestaltism," a term that we take up at the end of this subsection.

One of the most important publications of the Gestaltists is Max Wertheimer's (1923) paper "Untersuchungen zur Lehre von der Gestalt," where he described some Gestalt grouping principles. Three of these principles are demonstrated next.

The Principle of Proximity

The dots in Fig. 3.17 may be perceived immediately as arranged into three groups. Because they are all of equal size, form, and color, the only perceivable differences between two dots are their position and their distance to adjacent dots. Thus, proximity is the factor that is responsible for the emergence of the three groups.

FIG. 3.17 Grouping according to the Gestalt principle of proximity.

The Principle of Good Continuation

The arrangement of dots shown on the left-hand side of Fig. 3.18 gives rise to more than one possible interpretation. They may be seen, for instance, as a V-like figure that touches its mirrorimage below, as a conic figure on the one side touching its mirror image on the other side, or as two curved lines crossing each other. The latter is the organization that most perceivers characterize as the dominant one.

The central dot is the critical element of the figure. The V-like and the conic figure contain an abrupt change of direction at the central dot, a *discontinuity*, whereas the curved figures do not. Apparently, our visual system tends to avoid perceiving discontinuities in direction. This principle has been called the *principle of good continuation.*

FIG. 3.18 Grouping according to the Gestalt principle of good continuation.

The Principle of Similarity

The arrangements of dots and squares shown in Fig. 3.19 seems organizable in several equally salient ways, if one takes into account only the principle of good continuation. Organizations into horizontal rows, vertical columns, and diagonals are possible. Yet in both figures there is only one dominant arrangement: The left figure can be seen as organized into rows and the right figure may be perceived preferably as organized into columns.

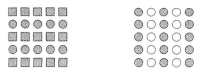

FIG. 3.19 Grouping according to the principle of similarity by form and color.

This effect is due to another Gestalt principle, that of *similarity*. The rows in the left figure contain only elements of similar form, and the colums in the right figure contain only elements of similar color. Grouping by similarity may be achieved as well with other features, such as similar orientation, brightness, and size.

The Gestalt principles provide a set of rules of thumb for deciding which elements of a picture go together with which other ones. Because this decision is an elementary problem of computer vision research, there has long been interest in using implementations of the Gestalt principles for image understanding tasks. A prerequisite for this is a theory that provides a basis for the quantification of the rather qualitative Gestalt principles. It was not until the 1980s that promising theories for some aspects of Gestalt perception came up. Because then, the number of publications on computer simulations of Gestalt grouping has grown remarkably and there has been renewed interest in Gestalt psychology: Rock and Palmer (1990) recently reported the discovery of some to date unknown Gestalt principles.

State of the Art

The idea of using simulations of Gestalt perception to improve the quality of image analyzing systems is not new: One early work on the subject by Guiliano et al. dates back to 1961. At that point a theoretical basis for the quantification of the Gestalt principles was still missing.

Today there is a variety of systems and approaches to the simulation of Gestalt grouping. Gillies and Khan (1992) used an adaptation of the well-known *Hough transform*, a method that is used in many computer vision systems. Kahn et al. (1990) and McCafferty (1990) used energy-minimizing algorithms, e.g., *simulated annealing*, to simulate Gestalt grouping, whereas Rosenfeld (1986) proposed the use of *pyramid algorithms*—another well-known architecture used in image recognition systems—for perceptual grouping of picture elements.

Almost all of these methods are not suited for the kind of application we have had in mind, graphical search, because they lack some important properties: Most systems perform grouping, but not recognition, and many systems are parametric; that is, the quality of the found groupings depends on some interaction with a human operator. Another problem with existing systems is the different types of data they process: Most of the systems are applied either to abstract dot patterns or to pictorial raster images, whereas the Graphics Designer's domain is object-oriented presentation graphics.

Our own system, MAX, is designed to find relevant groups without interaction with a user, and to provide a method for retrieving graphics from a database by matching groups in a search sketch and groups in the database.

MAX: Simulating Gestalt Grouping and Recognition

The implementation of the MAX grouping algorithms is based on Treisman's *feature map model* of human preattentive vision (Treisman, 1985) and on certain ideas from Palmer's *transformational model* of visual perception

(Palmer, 1983). Some key aspects of Treisman's model will be described in the next section.

The overall procedure for the simulation of Gestalt grouping of a presentation graphics is organized roughly according to the following steps.

1. Grouping by similarity of form, color, brightness, size, and orientation: For each of these features put each constituent graphical object into one similarity class according to its actual values regarding the respective feature. That is, each object belongs to five similarity classes.

2. For each object in each similarity class perform the following steps:

   a. Proximity grouping: Construct a local environment with a fixed radius that contains a maximum of eight nearest neighbors.

   b. Continuity grouping: Construct good continuations between neighboring local environments (this idea stems from Palmer, 1983).

3. Repeat step 2 for all objects that remained ungrouped, using a larger radius.

4. Stop if all objects are grouped or if the radius of the local environments has reached its maximum, that is, the maximum extent of the presentation graphics in any direction.

5. Collect the groupings found in all similarity classes, unite the results, if possible, and sort them according to some heuristic saliency criteria.

We explain these steps in a more detailed fashion in the next subsections.

Simulating Similarity Grouping

The simulation of similarity grouping is performed by MAX according to Treisman's (1985) feature map model of preattentive vision. In Treisman's model, a perceived picture is split up into a number of intermediate pictures along several feature dimensions. These intermediate pictures are called *feature maps*.

Each feature map consists of those parts of the picture that contain a certain feature. Feature maps preserve spatial relations roughly and may signal activity—that is, they may signal that there is a certain feature present in the scene, for instance, the color green, and how much of that feature is present. The integration of features to a complete picture is achieved by visual attention. Attention is directed to certain feature maps through a master map of locations that contains pointers to all the feature maps. There are two basic questions concerning this model:

1. What are the features that are relevant for human vision?

2. How are scene elements assigned to feature maps (how are feature values "coded")?

Treisman (1985) worked on both subjects and found some evidence that brightness, color, size, orientation, form, curvature, blobness, and line discontinuities may be relevant visual features. She hypothesized that feature values are coded relative to some normal value; this accounts for a number of known psychological constancy effects, such as *brightness constancy.*[7]

Treisman's feature maps may be conceived as similarity classes, so we have adapted the preattentive part of the model for our simulation of similarity grouping in MAX. During this simulation, all EPICT objects constituting a presentation graph are assigned to feature maps—at most five—in the feature dimensions of form, color, brightness, orientation, and size. In other words, if a graph contains only red and blue squares and circles of similar size, orientation, and brightness, then there is one feature map that contains all the circles, one that contains all the squares, one that contains all the red objects, and one that contains all the blue objects. Because the same object may be contained in several feature maps, it may also belong to several found groupings, if any.

Because constancy effects are of minor importance in the Graphics Designer's domain of presentation graphics, we have chosen an encoding that uses absolute feature values. The assignment of elements of a presentation graph to feature maps is performed in a heuristic manner, using quantizations of feature values, and according to some basic psychophysical laws. A detailed description can be found in Rome (1994).

After the feature maps are created, the simulation of proximity and continuity grouping takes place separately for each feature map.

Simulating Proximity Grouping

The simulation of proximity grouping is a preparatory step for continuity grouping, because good continuations are formed only between objects that have equal or similar distances. We have chosen a *multiresolution approach*

---

[7] Humans may perceive a color, such as white, independent of different lighting conditions. Lighting conditions may change in such a way that a formerly white region may turn to a light blue, "objectively"; still, this area may be perceived "subjectively" as white. This effect is known as brightness constancy.

(cf. Rosenfeld, 1986) to deal with different object distances and the overall procedure is as follows:

Each object of a feature map is assigned to a particular resolution level. We start with the highest resolution, that is, with the smallest distances. All objects on this resolution level are passed on to the continuity grouping procedure. If it leaves objects ungrouped, we use a lower resolution, that is, larger distances, and try again to form continuity groups. This procedure stops when all objects have been grouped or when the maximum possible distance has been reached.

Objects are assigned to a particular resolution level using a combination of two known methods for the determination of nearest neighbors. MAX constructs a *local neighborhood* for each object that contains at most eight nearest neighbors within a given radius. On each resolution level this radius is the minimum distance between two of the objects remaining to be grouped plus a tolerance of 30%. For each object, A, and for each object, B, in A's local neighborhood, an entry is created in the *relation matrix*, a two-dimensional array. The entry holds A's and B's identifier, the distance from A to B and the direction in which B lies relative to A, that is, an angle. The relation matrix is the data structure that is passed to the continuity grouping algorithm.

Simulating Continuity Grouping

The simulation of continuity grouping is implemented as a search of all possible continuous paths in a directed graph, represented by the relation matrix. The vertices of the graph are the objects of a particular feature map and a particular resolution level; the edges are imaginary connecting lines between adjacent objects. Each edge (A B) is labeled with the distance E_{AB} between the objects and the direction n of the connecting line. The graph may be cyclic and disconnected; initially, some roots are located as starting vertices. Continuous paths are then constructed according to the three conditions of a *continuity criterion*:

Let $P = (O_0, \ldots, O_l, A, B)$ be the path constructed so far and let $(A\ B\ E_{AB}\ n)$ be the last edge. If there are edges $(B\ C_1\ E_1\ k_1), \ldots, (B\ C_m\ E_m\ k_m)$ with directions $k_1, \ldots, k_m$ degrees, such that $C_1, \ldots, C_m$ are not in P, then take that edge $(B\ C_i\ E_i\ k_i)$ as next edge, where the angle difference $|n - k_i|$ is minimal and less then a maximum of 46 degrees.

Colloquially speaking, the first condition helps to prevent cycles in the path construction process; cycles may occur if there are arrangements of objects that form closed contours. The second condition means that the direction of the new edge has to be as close as possible to the direction of the previous edge, and the last condition states that angles larger than 45 degrees indicate a discontinuity of direction. This last condition is a sharp boundary and was introduced due to a missing notion of curvature in MAX.

Integration of Grouping Results

Because grouping takes place separately in a number of feature maps, some postprocessing is required to integrate the grouping results. We use some simple heuristics that have proven to be reasonably effective but that still could be improved.

Graphical objects may belong to more than one feature map, and thus may also belong to more than one grouping. If identical groupings are found in different feature maps, these groupings are united and annotated with the feature/value combinations that were responsible for the similarity grouping. If continuity groupings share coincident initial or final objects, then these groups are joined to longer, perhaps even closed contours.

The final step is the sorting of the remaining groups according to some heuristic saliency criteria. Longer continuity groupings are positioned at the beginning of the list of groups, because longer contours may be more salient than shorter ones. Groups that cover a large area are also positioned at the beginning, because large objects may be more salient than smaller ones.

This completes the description of the MAX grouping algorithms, and we now turn our attention to recognition tasks, especially to the recognition of similar forms.

Similarity of Form: The Power of Polygon Matching

The most important feature of a primitive object and of a group object is its form; this is another rather global Gestalt principle. Unfortunately, the mechanical detection of form similarity is also one of the most complex tasks. To date there is no satisfying or completely accepted theory of what visual form is. For the purpose of recognizing the forms of groups mechanically, we thus have to work with suitable simplifications of the notion of form.

There are two nontrivial aspects of the form of a group of objects that may be described by polygons. *Polygonal arrangements* of a group of graphical objects are formed by taking the centers of the group's constituent objects as vertices of a polygon, just as indicated in the curved arrangement of squares in the left-hand part of Fig. 3.20. *Silhouette polygons* are polygonal approximations of the silhouette of a group of graphical objects as shown in the right-hand part of Fig. 3.20.

Groups that are formed according to the principle of good continuation may be perceived as having similar arrangements, although the number of elements may be different and the elements themselves may have different colors, forms, sizes and the like; this can be verified by inspection of the three groups shown on the left-hand side of Fig. 3.20.

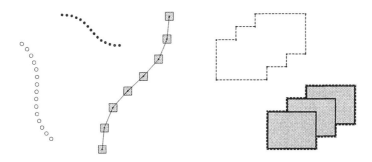

FIG. 3.20 Two aspects of form: arrangement and silhouette polygons.

Arrangement seems to be one important factor of the form of a group and we have concentrated on this issue. Other factors of form that are not handled by MAX are the silhouette of a group and the influence of the context, that is, of the arrangements of surrounding groups or objects.

Describing arrangements of continuity groups by polygons is only the first step to the computation of similarity of that aspect of form. We still need a procedure that determines whether two polygons are similar or not.

Arkin et al. (1989) developed such an algorithm that is capable of comparing and matching the forms of two polygons. It computes a similarity measure for the polygons that is invariant under translation, rotation, and dilation. Additionally, this measure has the properties of a metric and it is, to a certain extent, robust against "noise." The latter property means that similar polygons need not have the same number of vertices and that their edges may be slightly distorted.

This algorithm, which I refer to as the "Arkin metric," performs remarkably well. It has been integrated into MAX and serves several purposes:

1. It is used to compare the forms of arrangement polygons of groups that are formed by the simulation of the continuity grouping. That means:

    a. Similarity classes for the forms of arrangement polygons can be generated in the same fashion as for primitive polygons. This makes it possible to apply the MAX grouping procedure on its own results; that is to say, MAX can group objects in a hierarchical manner.

    b. The polygon matching algorithms can be applied directly by the Graphics Designer search component to compare groups in a search sketch and groups in its database.

2. It is used to compare the forms of primitive polygons in order to generate similarity classes for the form feature of this type of objects.

The Arkin algorithm takes two polygons and computes a numeric value that expresses the degree of their similarity. Still, we need some means to classify similar polygons in order to create feature maps for their form feature, that is, to create similarity classes. To achieve this, we have employed a nonparametric numerical clustering method developed by Umesh (1988).

Umesh clustering takes a number of data objects—polygons in our case—and a distance measure—here, the Arkin metric—and constructs plausible similarity classes of polygons automatically. No interaction with a user is required and no parametric tuning of the procedure is necessary.

One drawback of the original clustering algorithm is its inability to create rest clusters of objects that are dissimilar to all other objects under consideration. In our case, we could overcome this drawback by introducing a few modifications to Umesh's algorithm that exploited certain properties of the Arkin metric. A detailed description of the entire matching and class building process may be found in Rome (1994).

Figure 3.21 shows the results of the classification of eleven different polygons, where polygons in one column belong to the same class. Four similarity classes of polygons were created by the described method and two unclassifiable polygons (rightmost column) remained. Please note that only the form of the polygons was compared, not other features, like fill pattern or color.

FIG. 3.21 Four similarity classes of polygons and two unclassifiable polygons.

Test Results and Performance

MAX has been tested with a large number of artificial test patterns and with a number of presentation graphics from the Graphics Designer database. For instance, the patterns shown in Fig. 3.17 through Fig. 3.19 are all grouped as described in the corresponding text.

One of the presentation graphics analyzed by MAX is the graph shown in Fig. 3.16. We used two versions of the graph for our tests: In version 1, the shadowed text boxes were represented as complex, grouped objects, and in version 2 we used only primitive objects—that is, we partitioned the shadowed text boxes into their constituent rectangles and text objects.

In version 1, MAX found the following groups, with kinds of grouping indicated by initials [continuity (c), similarity of form (sf), color (sc), size (ss), etc.]:

1. The diagonal arrangement of text boxes (c).

2. The frame rectangle and a white rectangle covering part of it in the lower left corner (c, sf).

3. The title line "Assisting Computer (AC)" and the text line "Hoschka AC 23" (c, sc).

4. The text line "GMD Institute..." and the text line "Hoschka AC 23" (c, sf).

5. The text lines "Intelligent..." and "Planning...", and five other group-ings that are less salient.

In version 2, MAX found the following groups:

1. The diagonal arrangement of white rectangles (c, sc, sf).

2. The diagonal arrangement of text objects inside the white rectangles (c, sc, sf).

3. Four two-element groups, each consisting of a shadow rectangle and the text object in the corresponding white rectangle (c, sc).

4. Four two-element groups, each consisting of a white rectangle and its shadow rectangle (c, sf).

5. The other groups were like the ones found in version 1.

By hierarchical grouping, the arrangement polygons of the continuity groups 1 to 4 found in version 2 can be combined to yield a global group of "shaded text boxes": The arrangement polygon of the diagonal of white rectangles, that of the black rectangles, and that of the text blocks can be grouped by similarity of form and orientation.

This is an example showing that MAX is able to find salient groupings of graphical objects. The fact that it also finds less salient groups shows that the postprocessing stage may still be improved. Several researchers have already attempted to implement some "goodness criteria" to rate groupings, thus attempting to implement the global Gestalt principle of *figural goodness*. Such criteria are hard to implement for two reasons: (a) The decision which group is good and which is bad depends on the observer, and there are, of course, large differences between individuals and (b) there is as yet no accepted theoretical model for the respective Gestalt principle.

Conclusion

We have introduced a system, MAX, that is able to find salient nonlocal structures in presentation graphics that are represented in an object-oriented fashion. The method used is the simulation of the Gestalt principles of human perceptual organization. Salient nonlocal structures, that is, Gestalt group-ings, serve as additional information for other components of the Graphics Designer system, because these structures are not explicitly represented.

The graphical search component may use structures contained in a graph as indices for the location of that graph in its database. The retrieval process may also be supported by MAX: Structures in a search sketch supplied by a user may be compared directly with structures in the database using a polygon matching algorithm. Although MAX was not integrated into the graphical search component throughout the duration of the project, we have shown that the technical requirements were adequately met.

The implementation of MAX may still be improved technically up to a certain extent. But there are limitations in the underlying theoretical basis that can only be overcome by improved theories and models. Particularly desirable are the development of better models of visual perception and of models of figural goodness.

As far as graphical search is concerned, the approach of content-based retrieval of pictures and graphics is an up-to-date research subject (Barber et al., 1993; Kato et al., 1992). In order to improve the assisting capabilites of such systems, cognitively oriented methods such as MAX seem promising. The further development of such methods requires additional research effort in particular areas, such as psychology. Little is known about the question as to which features of a picture or a graphic are remembered well and which are not. Respective results may help, for instance, to develop heuristic goodness criteria to identify salient groupings more reliably.

3.5 INTELLIGENT DESIGN ASSISTANT

Angi Voß, Wolfgang Gräther, Jörg W. Schaaf

Current expert system technologies reach their limits in ill-structured domains, where there are more exceptions than rules and where many facts have to be qualified. Such domains are more widespread than was believed in the boom years of expert systems. Schools do not emit experts, but years of practice do. Experience is not well captured by a neat set of inferences, but rather by a collection of episodes or cases that may be heterogeneous, inconsistent, and fragmentary. Moreover, humans are not perfect in memorizing. They tend to forget some cases or remember them only vaguely. Even worse, the cases that make up one's experience are personal and cannot be shared as such. When the cases are externalized and stored in a public library, other users do not have the cues to access the right cases at the right time—you cannot remember what you do not know.

Here case-based support systems come into play. They maintain an episodic memory or, technically spoken, a case base and provide methods for storing, indexing, and sometimes for forgetting cases. Given a particular working context or problem, they will search for cases of similar previous situations. The simpler systems just return some cases, whereas more advanced ones can do a comparative assessment of the current problem, or even help to solve it by adapting solutions from the cases.

A case-based reasoner is much more than an information system: The cases have to be represented in a particular way that allows one to interpret them as problems and solutions. There is knowledge about similarity in the methods for indexing, storage, and retrieval. Assessment and adaptation methods incorporate knowledge about good and bad solutions and about how to match problems and how to transfer solutions analoguously.

Although conventional rule-based expert systems must be consistent and should be complete, case-based reasoners are more flexible, because the knowledge embodied in and distributed across the cases may be partial and partially inconsistent.

So far, case-based reasoners have mainly been used for classification and diagnosis, as decision support, and in help desk applications (Kolodner, 1993). Commercial case-based shells like Remind, CBR-Express, and

Esteem essentially represent cases as feature vectors and compare them by distance-based similarity functions. They support retrieval, but hardly assessment and adaptation, which require more domain-specific knowledge structures. Convincing concepts for integrating cases with more general assessment and adaptation knowledge are still to be sought.

This is where the FABEL project[8] wants to make its contribution. It will develop a methodology for building case-based design support systems, as well as tools and reusable modules for the various subtasks. Initially the project concentrated on the retrieval of cases. Later we worked on assessment and adaptation. We built a prototype system that was demonstrated at CeBIT fair 1994 in Hannover, Germany.

We have committed ourselves to architecture as a concrete domain, more precisely, to the design and layout of buildings with complex installations. But we aim to cover a broader spectrum of technical layouts. Our approach should be particularly relevant for one-of-a-kind products like laboratories, schools, plants, ships, and so on, where a feasible solution within an acceptable amount of time is preferred to an optimal solution obtained after extended design periods.

In spite of different baselines, namely, the integration of cases with general design knowledge in FABEL versus coping with vague knowledge in TASSO, both projects share some aspects (see chapter 2, section 2.3):

1. The vagueness of a current situation (TASSO) can be reduced by transferring information from similar, but more elaborated situations (FABEL).

2. Business graphics and the layout of office rooms (TASSO) are close enough to the layouts of buildings (FABEL) for the techniques applicable for one domain to carry over to the other.

3. Indeed, both projects investigated a quasi-analogous interpretation of layouts in terms of bitmaps.

4. The associative memory developed in TASSO has been reused in FABEL.

---

[8] This research was supported by the German Ministry for Research and Technology (BMFT) within the joint project FABEL under contract number 01IW104. Project partners in FABEL are German National Research Center for Computer Science (GMD), Sankt Augustin, BSR Consulting GmbH, München, Technical University of Dresden, HTWK Leipzig, University of Freiburg, and University of Karlsruhe.

5. Finally, there are similar ideas of identifying holistic gestalten or composite patterns in business graphics (system MAX, see section 3.4) as well as in architectural layouts.

This chapter concentrates on the retrieval of layouts in FABEL. We start with a short introduction to our domain, the design of buildings with complex installations. We show how an incomplete layout can be used as a vague query, and how the layouts retrieved can help to complete, assess, or modify the current problem. We extend the indexing scheme to incorporate complex gestalten in order to improve the precision of the method.

Building Design

An Integrated Building Model

Computer support essentially depends on suitable representations. Ideally, one would like to have one data model integrating all aspects of building design. We found such a model in the approach by Haller and Hovestadt. Haller developed three component-based building systems, among them MIDI for multistory buildings with complex installations like schools, labs, or office buildings (Haller, 1974). He equipped it with a generic installation methodology called ARMILLA to cope with the complexity of integrating the different subsystems [Haller 1985]. Hovestadt implemented a sequence of experimental software prototypes A1-A4 for computer-aided design of MIDI buildings according to the ARMILLA scheme. A4 has matured to a general integrated data model for buildings that is no longer specialized to the Haller school (Hovestadt, 1994). A4 building models are presented and manipulated by a system called DANCER.

The design of a building and the control of its life cycle rely on different media: text, graphics, sketches, sounds, videos, tables, faxes, and others. A4 encapsulates these multimedia sources in data containers that are to be manipulated by special purpose software. A4 itself allows one to create and relate such containers, called A4 objects, and to locate them in a multi-dimensional design space that encorporates a global coordinate system as a subspace. The dimensions most relevant for FABEL are:

Location and extension in x, y, z.

Aspect: subsystems like construction, fresh-air system, return-air system, or electricity.

Morphology: for functional, logical, and structural descriptions like usage or equipment.

Precision: zones (reserving maximal space, represented as ellipses), bounding boxes, elements.

Scale: orders of magnitude with 0 indicating the level of the whole building and 8 indicating parts of a room.

The innovative idea of A4 is to add semantic information not as separate attributes, which might even be stored in a separate database, but rather as additional dimensions. Thus, the designer can use the familiar mechanisms for orientation and navigation in three-dimensional space to inspect the abstract semantic information. This is crucial, because a building model may consist of tens of thousands of A4 objects. At any time, the architect or civil engineer will be working in a limited local area, an A4 subspace or view that he or she defines by restricting the values of some axes. Such a view typically contains from ten up to a hundred A4 objects. Views best compare to the layers of a more conventional computer-assisted design (CAD) plan.

The A4 subspaces may contain not only passive A4 objects but active agents as well. These can be architects or engineers operating concurrently, or assistance functions to assess or correct pieces of a layout. Such functions are typically concerned with one or two subsystems, morphologies, or scales; that is, they naturally associate with well limited subspaces.

The cooperation of these distributed agents is governed by a single principle: Only objects or agents that share some common space can communicate or affect each other.

Enhancing DANCER with Case-Based Retrieval

With DANCER as a drawing tool and editor for A4 building models, designing means creating, specifying, deleting, and arranging of A4 objects. When working with DANCER, one produces various views or plans of a building. They can easily be marked as cases by the architect. Thus, during the design of a building, the architect can acquire a substantial case base or extend already existing case bases.

Cases allow one to shortcut the design process, because solutions to problems do not have to be constructed from scratch but can be adapted from a case. Imagine the architect has produced a partial plan when he or she suddenly encounters a problem. The architect would like to look for similar plans that show how to detail the current plan, or perhaps how to modify it in order to avoid some difficulties. Defining queries is straightforward and naturally integrated in the architect's workflow: He or she selects some A4 objects in the current plan as the query and triggers the FABEL IDEA server by specifying a particular retrieval method, a case base, and maybe some additional parameters. The method looks for similar plans in the specified case base and presents them as thumbnail sketches. Selected sketches can be pasted into the current plan.

Figure 3.22 shows an DANCER screen with a large plan of the return and fresh-air systems at one floor in the MURTEN building, the education center of the Swiss railway company. The right bottom corner of the plan is magnified in the window on the right upper corner. It constitutes the current context. The marked objects represent the query. The window with the unicorn activates the FABEL IDEA server with the ASM method to be described next. The window below contains the thumbnail sketches of similar plans found by the server.

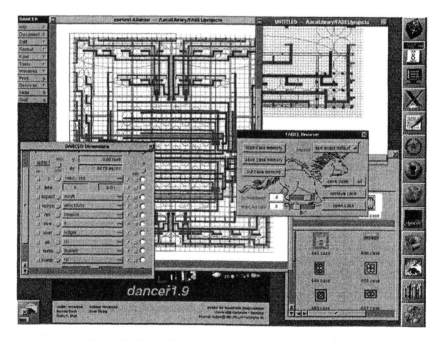

FIG. 3.22 The A4/DANCER-FABEL IDEA user interface.

How to Ask for the Right Information—Supporting Vague Queries

Because two cases or plans are similar if their indices match, retrieval involves indexing the query, searching the memory, and comparing the indices. Retrieval should be fast, a few seconds at most, so we should minimize the time for all three subtasks.

For memory search and comparison we have chosen an associative memory (ASM), which is a specialized two-layered neuronal network. Its input nodes correspond to individual features or attribute values; its output nodes correspond to cases or plans. The memory directly associates a feature with the cases possessing that feature. Memory search and index comparison are not conceptually distinct phases. The query indices directly activate cases with those indices. Only the relationship between query indices and case indices can be chosen. ASM is fast. It guarantees response times below a second for over a hundred cases, and it will scale up.

An A4 plan is a set of A4 objects, which are multidimensional vectors. Indices that could be obtained by condensing these object representations into a single vector for the whole plan would be fast to compute. So the questions were, what dimensions to be ignored and how do you merge the object vectors? You always have a trade-off between recall and precision. The more abstract the index, the better is recall, because there is less danger of missing a relevant case. The more concrete the index, the better is precision, because there is less danger of getting irrelevant cases. Finally, the more abstract the index, the easier it is to compute and the smaller is the case memory.

We opted for high recall rather than high precision, expecting to couple ASM with more precise but slower retrieval methods. We decided to ignore the location and extension dimensions of A4 objects. The results are object types that differ in their values in the dimensions aspect, morphology, precision, and scale. Our experiments have shown that scale is quite a good substitute for the exact geometric dimensions. As a consequence of ignoring position, we had to ignore any spatial relations between the objects, which is bad for precision. On the other hand, the indices are invariant with respect to translation, rotation, reflection, and resizing, which is good for recall. We then counted the number of objects of each type. We called such indices exact or precise. For instance, the query in Fig. 3.23 has one exact index, namely (4 return-air outlet zone 4), and the query in Fig. 3.25 has two of them, (1 room) and (6 tables).

Alternatively, and introducing a further abstraction, we stuck to the object types, but classified the counts into a few, some, medium, many, and very many, yielding so-called vague or fuzzy indices. For instance, the query from Fig. 3.23 has (few return-air outlet zone 4) as its only vague index, and the query from Fig. 3.25 has (few room) and (some tables).

In our application, two kinds of relations between the indices of the query and the cases turned out to be useful:

- Match-by-containment: The query (index) is completely contained in the plan (index). That means the case may show how to elaborate the query.

- Match-by-overlap: Query and situation sufficiently overlap. That means the plan may show alternatives to the query.

Both kinds of matches are reflexive but not symmetric. Only containment is transitive. Both types of comparisons can be applied to the two kinds of indices, yielding two orthogonal parameters for ASM: containment or overlap of precise or vague indices. As we show in Table 3.1, they yield different cases and are suitable for different situations and purposes.

TABLE 3.1
Precision of Retrieval; Depending on the Indexing and Matching Modes of ASM

Match	Precise Indices	Vague Indices
By containment	High	Medium
By overlap	Medium	Low

Some Examples

The left portion of Fig. 3.23 shows an early stage in the planning of the air-conditioning system for a seminar building. You can see the facade, the selected return-air zones, and the five zones for the rooms in the first floor of this building. The architect uses match-by-overlap with fuzzy indices to look for similar plans. A retrieved plan is shown in the middle of the figure; it is from the same MURTEN building already occurring in Fig. 3.22. One sees that a building of that size would presumably need return-air zones for outlets in the middle of the building. With this hint the architect can modify the design as shown in the right-hand portion of the figure.

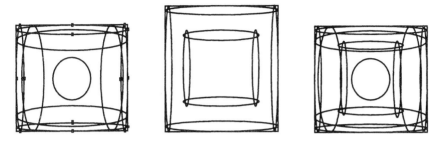

FIG. 3.23 A case for modification (query, retrieved plan, result).

In Fig. 3.24, left-hand side, the architect selects a return-air zone as the query ande would like to know what to do next. Both kinds of match with precise indices are the parameters for ASM retrieval. Four plans are returned; all of them show one return-air zone detailed with return-air outlets of scale 6. By using one of them the architect completes his design as shown in the right-hand portion of the figure.

FIG. 3.24 A case for elaboration (query, retrieved plans, result).

Figure 3.25 shows the seminar house in the stage of furnishing the rooms. The architect has put six tables in a room and selects them for a query with match-by-containment and exact indices. This means the returned cases must contain at least exactly one room and six tables. The thumbnail sketches for the retrieved cases are shown. They all solve layout problems.

Implementation and Evaluation

The associative memory was developed in the GMD project TASSO. It is implemented in Allegro CL (Henne, 1990). The retrieval method ASM is

FIG. 3.25 Another case for elaboration (query, retrieved plans, result).

built on top of it, it is also implemented in Allegro CL. A complete description is given in (Voß, 1994).

We experimented with three case bases. They contain from 42 to 126 cases covering different stages in the design process. The cases are layouts from the air-conditioning system, room layouts on floors, the construction of the building, and furnishing of rooms, for example. Our experiments showed:

- Retrieval time for match-by-containment depends on the length of the query (number of indices) and the connectivity of the associative network (case base); the time has always been less than a second on SUN Sparc ipx.

- Retrieval time for match-by-overlap is around one second in a case base with a hundred cases.

The number of precise and vague indices is proportional to the logarithm of the number of A4 objects in a plan. The size of the case base (associative memory) can be expressed by the number of nodes and the number of links between them. The number of nodes grows logarithmically with the number of cases; the number of links is proportional to the number of the indices of the cases. The case base can be incremented dynamically so that, on the fly, interesting (pieces of) plans can be marked as cases and from that moment on are available for retrieval.

Improving Precision by Recognizing Gestalten

The indices used in the associative memory are aimed at a high recall so that all relevant plans will be found. To increase precision and remove irrelevant plans, we have to take into account information about the spatial relations between the objects in a plan. However, computing all spatial relationships explicitly is very expensive, and still provides too low a level of information. Rather, we want to directly recognize complex constellations that "jump out and grab the architect". We now present a more knowledge-intensive method of how to grasp the main topological properties of a plan in very short time.

Gestalten in Architectural Layouts

Often constellations of objects remind architects or civil engineers of things out of their everyday lives like combs, fishbones, or quadrangles. Of course, the combs or the fishbones are not explicit in the plans. Because they are patterns in our mind only, we have chosen to call them gestalten, a term adopted from gestalt psychology (Helson, 1933; Köhler, 1929). We learned that architects use such "gestalten" to remember similar problems and their solutions. To elicit meaningful gestalten, we gave to our domain expert A4 plans, which he should associate with some tentative gestalten depicted on cards. He had additional empty cards on which he could draw new ones.

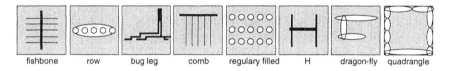

| fishbone | row | bug leg | comb | regulary filled | H | dragon-fly | quadrangle |

FIG. 3.26 The current set of predefined gestalten.

Figure 3.26 shows the gestalten that he repeatedly attached to the plans. We added indices like (3 row) or (5 quadrangle) as precise indices to the ASM retrieval method. Because usually there are only a few different occurrences of a single gestalt in a plan, we use one kind of vague indices only. For instance, (row) means one or more rows. The existing four alternative matching concepts still apply without any modification. In the FABEL IDEA server, gestalt indices still had to be attached manually to queries and plans by choosing them from a menu. Figures 3.27 and 3.28 show gestalt indices attached to an A4 plan and a query, respectively.

Sketching

Such manual indexing is not acceptable. It is too cumbersome in the face of some hundred gestalten for an entire building. Again, it was important to keep indexation time low, which meant we had to recognize the gestalten as fast as possible. The key idea was to find an abstract representation, or

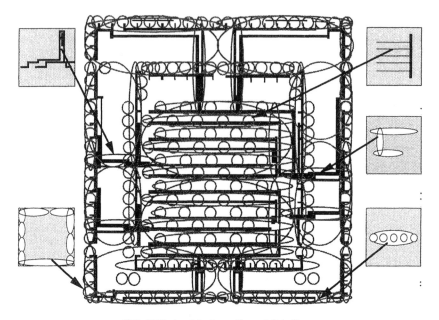

FIG. 3.27 An A4 plan with gestalt indices.

sketch, of the object groups in a plan. Examples of the gestalten would be sketched in the same way. Then an object group in a plan was said to have a certain gestalt if its sketch was identical to the sketch of some example of this gestalt. For the moment, let us assume we had some way to locate the object groups that constitute a gestalt. The questions then were what to represent in a sketch and what to neglect, so that comparison would be fast but correct.

FIG. 3.28 The situation from Fig. 3.25; but query with gestalt.

The main idea was to sketch a group of A4 objects in the same manner as a stick person sketches the constellation of head, body, arms, and legs of a human. The challenge was to find for each gestalt a single sketch that would represent all examples of this gestalt and none of the others, ideally. Figure 3.29 shows the origin of the idea and how it was elaborated.

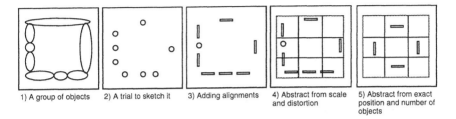

1) A group of objects   2) A trial to sketch it   3) Adding alignments   4) Abstract from scale and distortion   5) Abstract from exact position and number of objects

FIG. 3.29   How a group of objects can be sketched.

Focus on an object group: Part 1 of the figure shows an example of an object group that is often found in A4 plans. It may represent a way to divide the boundaries of a building into rooms and floors. Architects sometimes call this constellation a quadrangle. Each ellipse stands for the bounding box, here a room.

Try to sketch it: The second panel of the figure shows how a sketch would look like if only the centers of objects were represented. The problems are that on the one hand, too much information would get lost and, on the other hand, we would get different sketches for constellations of the same gestalt (e.g., if two quadrangles differ in size).

Add orientations: To avoid too much loss of structure, we decided to take the orientation of an object into account. The orientation is represented as a short line of the same orientation. A small circle indicates a missing orientation (third panel).

Abstract from scale and distortion: If the scale and the distortion of a gestalt were considered, it would be impossible to classify gestalten correctly. Humans also neglect these attributes while sketching things and in recognizing gestalten. Panel 4 shows a sketch that abstracts from scale (by scaling a grid) and from distortion (by distorting the grid in the same way).

Neglect exact position and number of objects: In the previous sketches, the exact number of objects is noticed. To abstract from similarly oriented objects, we defined a way to merge objects into one element. Panel 5 shows the result of this merging.

Thus we constructed a sketch that may represent many different examples of the gestalt quadrangle. Examples of a sketche for each gestalt is given in Fig. 3.30.

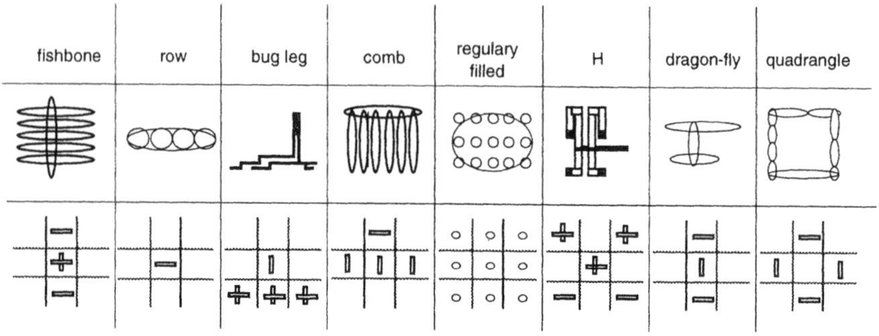

fishbone	row	bug leg	comb	regulary filled	H	dragon-fly	quadrangle

FIG. 3.30 Examples of object groups and their sketches

A first implementation of the sketching algorithm (in C++ on a SUN 4) shows that groups of one hundred objects can be sketched in less than a second. Gestalten are normally made up of a few dozen objects. Using real examples, we currently are experimenting with strategies for merging.

Next we would like to work out an additional top-down approach to recognize gestalten using the described sketching procedure. This method will try to make a gestalt presumption if the bottom-up approach produces sketches that do not match any of the predefined gestalt sketches. This presumption will be accepted if the complexity to repair the misbuilt sketch, to get the presumed one, is low.

Also, "object groups" have to be focused. Here the system MAX (Rome, 1993) developed within the GMD project TASSO, could be helpful. We would like to integrate gestalt recognition and focusing as an interdependent and interlocking process.

Outlook

In the preceding sections, we described and extended a fast method for retrieving layout fragments based on vague graphical queries. In the FABEL project, this method is supplemented with further techniques to compare layouts by their pixel images, or by graphs or terms describing the structure of objects using primitive operators or relations. The latter techniques perform an analysis that can be used to adapt the solution of a retrieved case to the current context. Other approaches to assessing and adapting layouts will be explored. One idea is to supply the A4 objects with knowledge about when their local environment is correct, and with operations to improve their local context. Another idea would be to parameterize the gestalten and to adapt the parameters to the current context.

The cases that fill our case libraries can be acquired both manually and automatically. At any time, the designer can select a set of A4 objects on the screen and create a case with them. The hand-selected cases will be special ones, while more routine cases are obtained from a method that extracts certain predefined views from a building model and turns them into cases.

Acknowledgment

We thank our colleagues in the FABEL project, Carl-Helmut Coulon, Friedrich Gebhardt, Barbara Schmidt-Belz, and Jürgen Walther, for their support in developing the ideas presented here.

3.6 USER INTERFACE DESIGN ASSISTANT

Harald Reiterer

Empirical surveys have shown that an average of 48% of the code in new business applications with graphical user interfaces (GUIs) is devoted to the user interface portion (Myers & Rosson 1992). Of the user interfaces of new applications, 74% were implemented using a tool kit, an interface builder, or a user interface management system (UIMS), and 26% were implemented using no tools. These results show that most software development projects spend significant time and resources on designing and programming the user interface and that most projects are using user interface development tools. The use of such tools does not guarantee GUIs of high ergonomic quality.

To reach GUIs of high ergonomic quality standards (e.g., ISO 9241; EN 29241), style guides (e.g., IBM, 1992; OSF, 1992; Microsoft, 1992) and guidelines (e.g., Mayhew, 1992) based on human factors knowledge have been developed. The volume of available standards, style guides, and design guides is enormous. Closely related to this circumstance is the fact, that today user interface designers need more and more competence, knowledge, and experience to handle this enormous amount of information. For many user interface designers this means that the execution of their jobs requires taking into account far more information than they can possibly keep in mind or apply. Empirical results have shown that most of the software designers have no or only very limited knowledge about human factors (Molich & Nielsen, 1990; Beimel et al., 1992). Therefore most of them were not able to apply standards, style guides, or guidelines in the design process, even though they have information or get information about their existence. In Beimel et al. (1992), designers were asked what kind of support they prefer to overcome their lack of human factors expert knowledge. A large number of them said that they would prefer computer-based design aids that should be integrated in their development tools.

However, the problem is how to capture and encode design knowledge relevant to the designers' tasks and how to present it to them in formats that support their mode of work. The result is a need for user interface development tools with domain competence based on human factors knowledge (e.g., standards, style guides, design guides), which may be encountered, learned, practised, and extended during ongoing use—in other words, tools in which users learn on demand (Eisenberg & Fischer, 1993). An important research goal in the area of user interface development tools is therefore to discover helpful, unobtrusive, structured, and organized ways to integrate the use of principles, guidelines, standards, style guides, and design rules into the tools without stifling creativity (Hartson & Boehm-Davis, 1993). These research issues should include methods and tools for offering the designer assistance in understanding, searching, and applying design principles, guidelines, and standards. This leads to the questions, what is the best presentation format for communicating user interface design knowledge, and how could we ensure that it will be observed?

Design Aid Tools for Communicating User Interface Design Knowledge

The development of design aid tools for offering the user interface designer assistance in understanding, searching and applying ergonomic design knowledge was the starting point for the GMD project IDA (User Interface Design Assistant) (Reiterer, 1993, 1994). The primary goal of this project is to incorporate domain competence in user interface development tools to empower the user interface designers. This means the development tools and the designers are bringing complementary strengths and weaknesses to the job. Rather than communicating with tools, the designers should perceive the use of the development tools as communication with an application domain (Fischer & Lemke, 1988). To shape the tools into a truly usable and useful medium, the tools should let the designers work directly on their problems and their tasks. The following design aid tools assisting the designers of GUIs during the design process have been developed in the IDA project:

- A library of reusable ergonomic interface software (construction tool).
- An adviser for ergonomic design support (advice tool).
- A tool evaluating the ergonomic quality of GUIs (quality assurance tool).

All design aid tools are integrated in a user interface management system (UIMS). The integration of the design aid tools into the UIMS is important. Separate systems for construction and enabling have major deficiencies. If the argumentation of the enabling system is to serve design, it must do so by informing construction. This can happen only if construction, argumentation, and evaluation are explicitly linked in an integrated design environment.

Presentation of the Design Aid Tools

To present the different design aid tools to the designer in an integrated
fashion, a control panel called IDA-Toolbar has been developed, which is
permanently placed on the screen. Figure 3.31 shows the IDA control panel
placed over the UIMS and a small example of a GUI, which is shown under
the UIMS. The designer can communicate with the IDA design aid tools by
the help of the control panel and its icons. If the designer wants some advice
during the design of a GUI, he or she activates the relevant icon in the IDA
control panel and gets a global or context-sensitive support through the
selected design aid tool. A control program behind the IDA control panel
controls the communication and the data flow between the UIMS and all the
design aid tools. This control program defines a clear interface between the
different UIMS and the design aid tools. The control program is also responsible for the coordination of the different design aid tools. All project results
have been developed under Motif/UNIX and Windows/DOS.

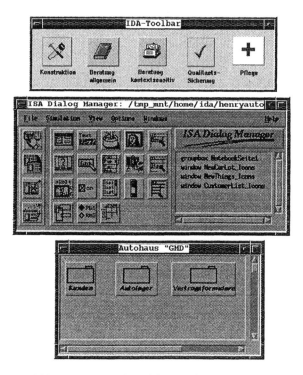

FIG. 3.31 Presentation of the IDA design aid tools.

An important requirement for the project was to develop prototypes of tool-independent design aid tools. In principle it should be possible to integrate the features of the design aid tools in each commercially available UIMS. Therefore different UIMS have been used as platforms for the integration of the design aid tools. It is the task of the manufacturers of the UIMS to integrate the features of the different design aid tools into their UIMS.

Support During the Construction of the User Interface

The IDA construction tool offers domain-oriented templates in a library, such as generic and domain-specific interface objects ("look") and dialogue scripts ("feel"). Based on object-oriented mechanisms of the UIMS, templates of generic and domain-oriented building blocks are constructed—under consideration of the relevant ergonomic standards, style guides and guidelines—and saved as object classes in the library. Using the library the designer generates an instance from each template. This instance will be integrated in the interface under design. Therefore designers construct interfaces by obtaining predefined templates from the library and placing them into the working area of the UIMS. Now the designer can modify the instance of the template, based on specific application requirements. This allows a "design by modification approach."

An important requirement for the usefulness of the library is the available information retrieval mechanism. Without powerful information retrieval features the designer won't be able to find any relevant template for the specific design situation. Different information retrieval mechanisms have been built in the construction tool to assist the designer during the search for relevant interaction objects. One information retrieval mechanism integrated in the IDA construction tool is a graphical browser, shown in Fig. 3.32.

The browser is based on a semantic tree, structuring the templates. This structure differentiates the templates from a designer's point of view and is based on typical tables of content that one can find in style guides or guidelines. The designer has the possibility to browse through the library by using the semantic tree structure to find a relevant interaction object or dialogue script. If the designer prefers an alphabetic presentation of the predefined templates, there is the possibility of changing to an alphabetic listing of all predefined templates of the library. The designer can now browse through the listing or can use a query function to find a template.

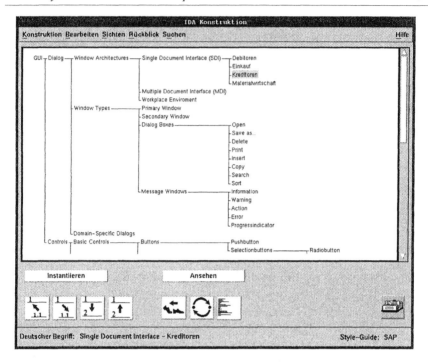

FIG. 3.32 IDA construction tool—graphical browser.

The query function is another information retrieval mechanism based on key-words (e.g., typical use of the templates, typical attributes of the templates, names of the templates) to search for a specific template. In a dialogue box, typical keywords are presented in a drop-down combination box to the designer, who can select one of the predefined keywords or use self-defined ones. The results of the search are shown as items in a list box. That allows a specific selection of templates.

If the designer wants to see the look and feel of a template before making a selection, he or she can start a simulation mode that shows the look and feel of the selected template. A visualization of the template, a list of all dialogue scripts attached to this template, and important attributes of the template are presented in separate windows. Figure 3.33 shows an example.

In the left window (entitled "<Arbeitsgebiet> <Operation>") the look and feel of a template for typical data entry purposes is shown. In the right window (entitled "Regeln – Funktionen – Attribute") the attached dialogue scripts, application functions, and important attributes are shown. The

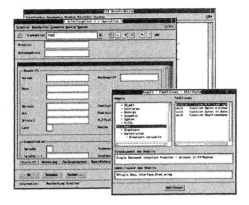

FIG. 3.33 IDA construction tool—look and feel of a template.

designer can now explore all important characteristics of the template, such as its look and feel, because the presented visualization can be used in the same way as the instance of the template. If the designer has found a template with the relevant look and feel, he or she can make a sample of the template. This sample will appear in the working area of the UIMS and can then be used in the construction process of the user interface. If the designer needs further information when and how to use a specific template, he or she can activate the context-sensitive advice tool with the help of the advice icon ▣ (see Fig. 3.32).

The IDA construction tool does not help designers to perceive the shortcomings of an interface they are constructing; it includes only passive representatives. Constructions of user interfaces in the work area of the UIMS do not talk back unless the designer has the skill and experience to form new appreciations and understandings when constructing. Designers who are unaware of human factors knowledge about user interface design do not perceive a breakdown if one of the design guidelines is violated. To support designers in this situation, two other design aid tools were developed in the IDA project.

Advice During the Design Process of the User Interface

The IDA advice tool assists the designer during the design of the user interface. If the designer needs support in the area of ergonomic user interface design, there is global or context-sensitive advice through activating the advice tool. The aim of the advice system is to determine an analogy between the examples of the adviser and the current task of the designer. The designers are aided in building an analogy by assuming that the presented example or information is relevant to their current task. If the designer wants deeper information—for example, why the "look" and "feel" of a specific interaction object should be designed in a certain way—the designer gets it in a form of a hypertext advise based on multimedia documents that are developed with the help of multimedia tools.

An important aspect in multimedia systems is the navigation support to prevent users from getting "lost in hyperspace." Therefore different powerful navigation mechanisms have been integrated in the advice tool. A graphical browser, an index, and a text query function are available as global navigation tools. The graphical browser is based on the same structure as the browser of the construction tool and structures the design knowledge in different sections. With the help of this browser the designer can browse through all sections of the advice tool searching for an interesting topic. The index and the text query function allow direct access to interesting topics.

A different concept to present design knowledge in a global way is the guided tour of user interface design. The purpose of this guided tour is to explain how graphical user interfaces should be designed considering human factors. Figure 3.34 shows the start window of the guided tour. Based on an object-oriented GUI development life cycle—consisting of six steps—all necessary design activities are described.

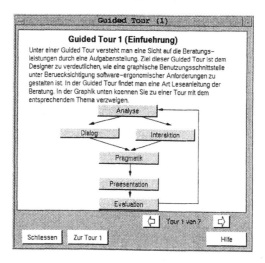

FIG. 3.34 IDA advice tool—guided tour of user interface design.

For each development step a predefined sequence of sections that the designer should follow is established. If the designer goes through all six development steps; he or she will have visited all sections of the global advice. This form of presentation allows a guided use of the advice tool and is useful for the novice designer in the area of GUI design.

Activating the context-sensitive advice, the designer gets some information on when and how to use the selected interaction object in the UIMS and what should be the ergonomic "look" and "feel" of this interaction object. The selected object could be a generic interaction object (e.g., push button, list box, edit text), a sample of a template, or a part of a complex dialogue object. The necessary context information is transferred from the UIMS to the advice tool to select the relevant section of the multimedia advice system. After activating the advice tool the designer gets a short definition of important features of the selected interaction object. Figure 3.35 shows a typical example of a multimedia document. Each document includes different navigation controls (menu items, icons, push buttons, hot spots), which allow a differentiated presentation of knowledge.

If the designer wants a detailed presentation of all relevant design guidelines for this interaction object, it is possible to open a follow-up window with the help of the push button "Richtlinien" (see Fig. 3.35). Figure 3.36 shows the related guideline window of Fig. 3.35. It offers in a textual and graphical form an explanation of all important guidelines that have to be considered in using or designing this interaction object. Different colors are used to distinguish the different sources of guidelines (e.g., style guide, standard, literature). The guideline text on the right side will appear after clicking in the graphical example on the left side. Depending on the selected area, the relevant guidelines appear.

If the designer needs further information on when and how to use an interaction object, there is specific support with the help of the push-button "Anwendung" (see Fig. 3.35). In a new window a question and answer dialogue appears. Depending on the answers, the designer gets recommendations about the use of available interaction objects and templates. In the main window of the advice tool the recommended interaction object or templates are shown.

For some interaction objects an animated presentation of its use is available. The designer could activate the animation with the help of the push button "Animation" (see Fig. 3.35) getting an explanation of the use and the behavior of the interaction object. The animation is developed as a small movie

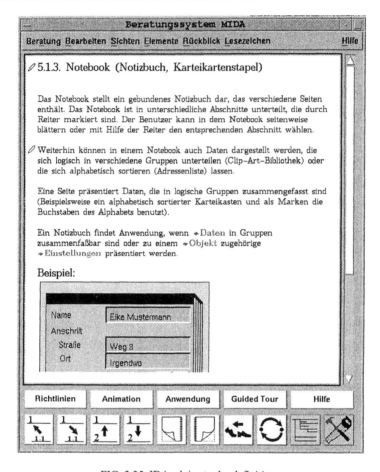

FIG. 3.35 IDA advice tool—definition.

that can be played, stopped, rewound, and so on, using a video player control panel.

Whenever a relevant template in the library of the IDA construction tool is available, the designer can retrieve it from the multimedia documents activating the construction icon ✖ (see Fig. 3.35). The instance of this template is automatically placed in the working area of the UIMS. Thus the designer always has the possibility of switching between the advice and the construction tool, which gives a maximum of flexibility in using the design aid tools.

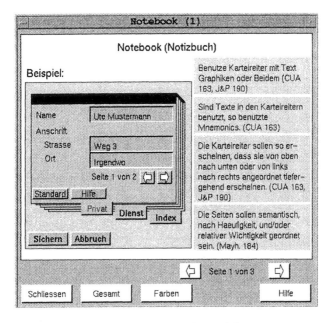

FIG. 3.36 IDA advice tool—guidelines.

Quality Control of the User Interface Design Results

The IDA quality assurance tool identifies potential problems in the artifact being designed. It detects and critiques partial solutions constructed by the designer. These critiques are based on human factors knowledge and design principles for user interface design. It is a form of performance critics whose prime objective is to help users create high-quality products in the least amount of time using as few resources as possible (Fischer et al., 1991). Learning is not the primary concern of the performance critic but may occur as a by-product of the interactions between users and critics. With the help of an expert system the knowledge is implemented in a knowledge base as interface objects and condition–action rules, which are tested whenever the designer asks for a quality control. A very similar approach is found in Löwgren and Nordquist (1992).

The passive quality assurance (critique) is explicitly invoked by the designer when desiring an evaluation. The result of the current design process is saved in a tool-dependent dialogue definition language (DDL) file of the UIMS. With the help of a parser this DDL file is translated in a tool-independent

DDL file. The expert system of the IDA quality assurance tool uses this file as input and represents the result of the dialogue design with the help of an object tree. Using the rules of the knowledge base and controlled by an inference mechanism the expert system analyses the conformance of the user interface with the human factors knowledge (analytic critique).This passive form of quality assurance usually evaluates the (partial) product of the design process, not the individual user actions that resulted in the product.

The results of the analytic critiquing process are presented in a special window shown in Fig. 3.37. The window of the explanation component contains a short description of all discovered ergonomic deficiencies (heading "ergonomische Abweichungen"), the source of the short description (heading "Quelle der Regel"), and the identifier of the evaluated object (heading "Instanz des Designs"), which contains some ergonomic deficiencies. Now the designer has different possibilities to continue the work. The designer who needs further explanation of the comments can activate the IDA advice tool with the help of the advice icon ▣. If a specific template is available, the designer can activate the IDA construction tool with the help of the construction icon ✖. If the designer selects the identifier of an object, he or she can activate the object editor of the UIMS (double-clicking the identifier) and can reimplement the interaction object that contains some ergonomic defi-

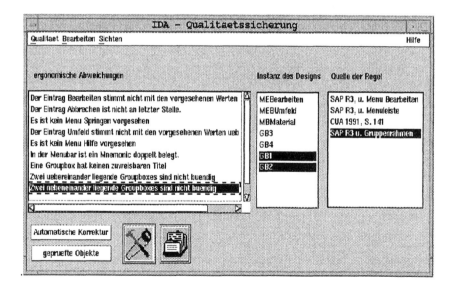

FIG. 3.37 IDA quality assurance tool—window showing ergonomic deficiencies.

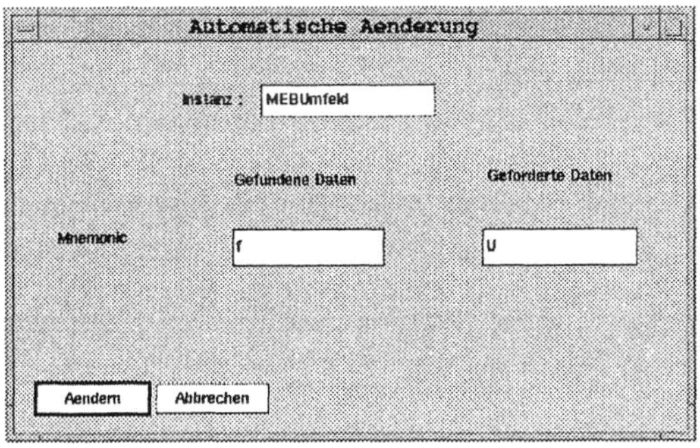

FIG. 3.38 IDA quality assurance—dialogue box for automatic correction.

ciencies. In some situations it is possible to change the detected deficiencies automatically. In this case the designer can activate a separate dialogue box (activating the button "Automatische Korrektur"), which gives the opportunity to correct the deficiency. Figure 3.38 shows an example of such a dialogue box that presents a faulty mnemonic of a menu item (labeled "Gefundene Daten") and the necessary correction (labeled "Geforderte Daten"). The benefit for the designer is in not having to branch into the editor of the UIMS reimplementing the mnemonic of the menu item.

The set of currently implemented rules covers different groups. The first group of rules represents user interface design principles of controls (e.g., list box, drop-down list, push-button, group box, entry field, check box, radio button). The second group of rules checks the content and structure of menus (e.g., menu bar, pop-up menu, pull-down menu, cascaded menu, mnemonics, short-cut keys). The third group of rules checks the general layout of a window (e.g., title, placement of group boxes, placement of push-buttons) and the layout of specific dialogue boxes (e.g., warning message, open dialogue, save as dialogue). The content of the actual rules is restricted to more static aspects of a GUI. It is intended to extend the rule base with rules that also covers the more dynamic aspects of a GUI (e.g., typical dialogue sequences) and with rules that represent application domain-specific knowledge (e.g., attachment of database functions to interface objects).

Conclusion and Outlook

Based on cooperations with scientific (Fachhochschule Darmstadt; University of Bonn) and industrial partners (Software AG, Darmstadt; SAP AG,

Walldorf; Hoechst AG, Frankfurt), prototypes of all design aid tools have been developed and connected with a commercial UIMS (ISA-Dialog Manager from ISA GmbH, Stuttgart). The usefulness and usability of these prototypes will now be evaluated from UI designers in the realistic context of their application domains. The following benefits for the designer using the UIDE are expected and will be evaluated:

- Designers will be able to learn human factors knowledge during their daily work using their development tool ("learning and use on demand").
- Designers will be enabled to apply ergonomic style guides and guidelines ("usability").
- Designers will be able to use predefined ergonomic interaction objects ("reusability").
- Designers will be able to evaluate the ergonomic quality of their design during the design process ("quality assurance").

All results of the evaluation process will be considered in the ongoing development process of the design aid tools to improve them.

A further intention of the ongoing development process is to build a specific design aid tool supporting the design decision process. With the help of such a tool the rationale for the various user interface design decisions can be made explicit and recorded for later reference. Having access to an audit trail through the design rationale is important during iterative development. Because changes to the interface will often have to be made, it is helpful to know the reasons underlying the original design. Design rationales can be captured in a hypertext structure or with the help of video records of design meetings and selected user tests. It is planned to extend the IDA advice tool with special features for capturing design rationales based on a special repository including hypertext documents and videos.

Acknowledgment

Several people have been involved in the IDA research project over the years. Thanks to Frank Bachmann, Doris Kamnitz-Kraft, Reinhard Oppermann, Michael Porschen, Manfred Ramm, Stefan Schäfer, and Helmut Simm for their contributions to the project results.

4 Systems to Support Cooperative Work

4.1 COORDINATION SUPPORT BY TASK MANAGEMENT

Thomas Kreifelts, Elke Hinrichs, Gerd Woetzel

Working in large and geographically distributed business organizations and government agencies requires that individuals and groups at different sites engage in intensive cooperative activity. Business teams are formed from different parts of an organization, and agencies in noncentralized government settings have to cooperate over large distances. There is a trend toward formation of joint ventures and consortia to carry out large projects across organizational and national boundaries. All this leads to an increasing need for computer tools to support distributed work management in order to provide a better overview and to avoid expensive inefficiency, errors, and delay.

Another incentive for the development of support tools for distributed work management is the emergence of alternative forms of the organization of labor: On the one hand there is an increasing number of people working "on the move," at a customer's or on a construction site. On the other hand, we observe a tendency toward people working at home in order to avoid commuter traffic and to enable more effective part-time working.

The increasing need for the management of distributed work makes the support of coordination an important assistance property that future computer systems should provide to their users. Consequently, the development of a flexible and integrated coordination support tool, which is to a high degree end-user controlled, was taken up in the Assisting Computer Project

(Hoschka, 1991). In the following, we report on the results of these research efforts.[1]

A Bit of History

Computers in the form of personal computers interconnected by computer networks have become a device used daily by many people in nearly all areas of business and at home. The role of the computer as a medium for communication opened up the possibility for new applications that no longer aimed at the exclusive support of a single user but rather at the support of a group of people. The development of this type of applications was also stimulated by the fact that people seldom do their work in isolation, but rather collaborate with other individuals in groups, teams, or organizations. This view of human work had altogether been neglected by single-user computer applications.

Although electronic mail provides a way of communicating via computer, it includes no support specific to coordination. Coordination support was first taken up by researchers in office information systems. They identified repetitive and structured office processes as the target for support. The goal of the various developments, starting around 1975, was creating means of specification and automation of business processes by providing an abstract model of business processes and supporting a whole class of processes covered by the model. Typical application domains were purchase procedures, business trip procedures, order processing, and loan application processing. The information dealt with usually was structured records, and the predominant paradigm was that of the "migrating form," which wanders through the organization displaying and collecting information at the various stations. The processes were then generally called office procedures, and the support systems were called office procedure systems; they are now more commonly known as workflow systems.

Today, there is a large number of workflow systems available on the market based on these early designs. Additionally, groupware tools have appeared that claim to support coordination in another way: Some cover certain aspects of the problem like personal productivity tools, group calendars for local networks, or project management systems (which actually are single-user applications for project managers); others represent closed "groupware" solutions or development platforms with no clear concept of, and not particu-

---

[1] Research reported in this chapter has been partly funded by the Commission of the European Communities in ESPRIT projects EuroCoOp and EuroCODE.

larly tuned to, coordination. The integration of, or interoperability with, existing computer support (like word processors, database systems, spreadsheets, etc.) usually presents a serious problem. On the market, there is not much support available that provides the comprehensive coordination functionality and the necessary amount of "openness" to third-party software that we consider essential.

A New Approach to Coordination Support

Within academic research, quite a number of models, prototypes, and systems have been developed with the explicit goal of coordination support, such as the AMIGO (Pankoke-Babatz, 1989) and COSMOS (Bowers & Churcher, 1988) projects. Not many of these approaches have ever been implemented or even put to use, so there is not much experience with coordination tools in spite of the obvious need for such tools. The few experiences reported on computer support for coordination exhibited a number of difficulties with those systems, particularly the lack of flexibility, low interoperability with existing tools, and a rather poor cost-benefit ratio. This also turned out to be true for the workflow system DOMINO that we had implemented (Kreifelts et al., 1984) and evaluated in a small field test (Kreifelts et al., 1991); the main problems we found can be summarized as follows:

- The rigidity of pre-defined procedures and imposed structures which lead to a limited application domain and to non-adequate exception handling in a number of situations.

- The isolation from informal communication, information sharing, and other forms of computer support.

- The effort needed to install and to use workflow technology before it pays off.

To avoid the implications of rigidity, it has been argued in Hennessy et al. (1992) that future coordination support systems should follow Schmidt's proposal (1991) of treating models of cooperative work as resources to be defined, modified, and referred to for information purposes, instead of as prescriptions to be adhered to. To overcome the relative isolation, future coordination support systems have to be able to interface to existing computer systems that support the actual work—coordination is never an end in itself.

The effort needed to make use of coordination systems is an aspect that seems to so far have been largely neglected: most systems require preorganization of the cooperative work by some sort of systems administrator before a system may be put to use by an ordinary user. Instead, one would prefer to have coordination systems that encourage self-organization of cooperative work by the end-users themselves. In order to overcome the initial barrier of using coordination systems, the genericity and simplicity of the underlying coordination model are of primary concern.

Within the Assisting Computer Project, we began to develop a new kind of coordination tool that was to address these design goals. The tool is called the Task Manager and is mainly meant to support day-to-day cooperative work in small- to medium-sized distributed teams.

The Task Manager

The Task Manager is a software system for specifying and managing cooperative activity in a community of geographically dispersed users. Coordination in the Task Manager is centered around the action-oriented aspects of cooperation and is based on a task-oriented coordination model (Hennessy et al., 1992). In modeling a cooperative activity, the tasks to be carried out by the individual group members are the central points of interest.

With the help of the Task Manager, users may organize (create, refine, and modify) cooperative tasks, monitor their progress, share documents and services, and exchange informal notes during task performance. The Task Manager distributes task specifications, attached documents, and notes to the involved users in a consistent way. It is meant to support the management of work distributed in time and/or space by providing

- Support of organization and planning of collaborative work (who does what, with whom, until when, using what material).
- Up-to-date overview of collaborative activity and work progress.
- Dynamic modification of work plans during performance.
- Availability and exchange of documents and informal notes within groups of people involved in task performance.

Typical examples of use are the preparation of a workshop or a project review, the collaborative generation of a report or proposal, and the development of a software system prototype. The Task Manager is to help organize such activities by supporting the planning as well as the performance, also under the condition of frequently changing goals and work assignments.

The Task Manager organizes distributed work in tasks that have a person responsible, a deadline, other participants, the material needed for task performance, and possibly subtasks. The primary user interface of the Task Manager is a hierarchically structured to-do list, which displays all tasks a user is involved in and which allows direct and easy access to the relevant information attached to a task.

Components of the Task Manager

The central notion of the Task Manager is that of a *task*. In order to perform a task, *people* use shared *documents* and/or *services* and communicate by exchanging *informal notes* with each other.

Tasks: The Task Manager's notion of a task has various aspects: one could think of a task as of a project, that is, a common goal of a set of people (*result-oriented*). A task may be broken up into several subtasks, and dependencies may be defined between them. The more detailed are the specifications given, the more a task resembles an office procedure with dependencies between subtasks and documents of a task (*procedure-oriented*). A task may also be used as a simple folder with little or no structure defined; in that case a task is simply a shared container of subtasks, documents and/or services, and notes that people exchange about in a common task (*information-sharing-oriented*).

Documents/services: In everyday office life, there are generally resources of some sort needed in order to achieve the goal of a task; therefore, each task may have resources attached to it. In our system, resources are "pointers" to various kinds of computerized objects; the first kind of resource one could think of would of course be documents that are shared between the participants of a task. But there are also other resources like rooms, budgets, machinery, and so forth that may be crucial to the performance of a common task. In our system, those resources are handled by services that are implemented outside of the actual Task Manager, but may be referred to and shared by the participants of a task.

People/users: The most important "resource" is the people involved in a common task. We distinguish between various levels of participation and of competence. There is a set of people involved in a task, the *participants*, that all have equal access rights to the attributes of a task, its documents and services and its notes. Participants may invite other people to take part in the task, that is, to become new participants or observers. *Observers* are people interested in the completion of the task with read access only to any information and the right to participate in the exchange of informal notes associated with the task.

One of the persons involved in a task is "more equal" than the others: the *person responsible* for the performance and the outcome of the task. This person has exclusive write access to some of the tasks attributes, such as state, start date and deadline, and is the only one who may reassign the responsibility of the task to another person.

Informal notes: Participants and observers may freely exchange informal notes within the context of a task. By integrating and facilitating extensive exchange of textual notes we give more room to flexible social protocols in contrast to regulations dictated by the system.

Tasks, participants, documents, and services are specified in more detail by a set of attributes each. The *title* of a task both identifies a task to its participants and gives a short and concise description of its goal. The title of a task is the only mandatory attribute—thus, the user is not forced to fill out long forms before actually starting work on a task. All other attributes are either optional or set to a default value by the system. For example, when a user initially creates a task he or she automatically is the person responsible for it.

As mentioned earlier, the person responsible for a task has special rights; in particular he or she may set the *state* of a task. We only distinguish between the completion state of *not finished* and *finished*. This state is set by the user. Apart from the completion state, a task can be *pending*, that is, it depends on another task not yet finished, or *not pending*. This state is set by the system automatically, and should help the user decide when best to start with the actual work on a task. As a third kind of state information, a task can be *acknowledged* by the responsible actor. This is to inform the coworkers of the responsible person's awareness and acceptance of the task he or she has been assigned to.

Another important attribute is the envisioned *deadline* of a task. The system reminds the user of approaching deadlines, but does not enforce any actions with respect to overdue tasks.

Besides those most important attributes, there are a number of other attributes that allow the user to specify in more detail how a task should be performed: timerelated data, such as start date, data that describe dependencies between tasks and between tasks and documents, and personal data attached to a task, such as personal files, priorities, etc. The latter attributes are purely local and are not distributed to and shared by the other participants.

Documents and/or services may be attached to any task in which the user participates at any time. They consist of a name, a history of who did what and when, the owner of the document, and other data. After discussions with prospective users, we added the *abstract* attribute of a document; it contains an informal text description of the document and it frees the user of having to transfer, open, and read the entire document when the user is only interested in a resumé. For reasons of simplicity, we decided to implement a semitransparent file transfer service (cf. subsection on implementation). Other resources, as already mentioned, are handled by separately implemented services; the system keeps an account of so-called service requests, but leaves other details to the respective service.

Participants are people that work together on a common task. They are worldwide and uniquely determined by a pair of IDs or by an X.500 distinguished name. We also support more user-friendly names, individual address books, and access to the X.500 directory service.

Functionality and User Interface of the Task Manager

Users can create tasks and subtasks, and dependencies between tasks and between tasks and documents; they may set and modify attributes, add, modify, and remove documents, and service requests. Persons responsible for a task may refuse responsibility and they may reassign it to another user. Any participant can introduce new participants or observers to a task. Tasks may be copied and pasted or moved around freely. All of this may be done at any time, thus allowing dynamic modification of the work situation at run time.

Basically, the system distributes information on tasks, makes available resources across the (worldwide) network, keeps the data up-to-date, and resolves conflicts of synchronization. Each user has instant access to the shared tasks he or she is involved in. The system guarantees a consistent view on tasks for each participant. Additionally, the system keeps track of the actions the users take. Thereby, monitoring and task tracking at execution time as well as report generation after completion of a task are rendered possible.

The set of tasks a user participates in is presented at user interface level as a Task List, very similar to conventional outliner programs (Fig. 4.1). The Task List gives an overview of the hierarchically ordered tasks along with a condensed view of the most important attributes: title, person responsible, deadline, documents/services, participants, state, and a list of notes that have been exchanged within that task. Operations are invoked by selecting menu commands and/or by directly typing in the attribute fields. In most cases, the Task List will suffice to display the information and to perform the necessary operations.

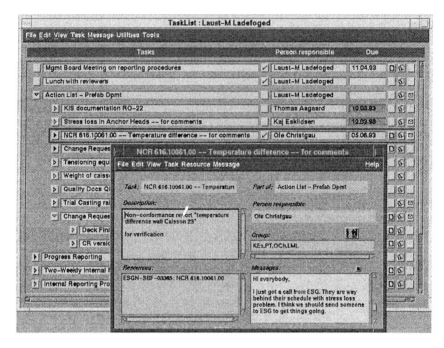

FIG. 4.1 Task List user interface.

For more detailed information, we provide a form-oriented Task Editor; it may be used to view and edit in detail all of the attributes (shown as an additional window in Fig. 4.1). The Task Editor also provides access to a local address book as well as to external addressing information such as the X.500 directory.

The Dependency Editor represents groups of subtasks and documents in a net-like way; it mainly serves the purpose of graphically displaying and editing dependencies between tasks and resources. The dependency structures are similar to those used in workflow systems, but are interpreted differently: instead of driving a workflow with a strict sequencing of steps, dependencies in the Task Manager represent recommendations that may be changed or overridden by the users at any time.

The Logging Tool gives a complete history of events with respect to all or to selected tasks, and lets the user acknowledge new events. Thereby, the user may find out what's new since "last Tuesday."

Other views are possible and desirable, but are not yet implemented: filters on the Task List on the basis of selection criteria, sorting according to other criteria, Gantt or milestone charts for a group of subtasks in order to facilitate time management, and so on.

Usage of the Task Manager

In this subsection we describe in more detail how we envisage the usage of the Task Manager. This also includes other forms of system usage not generally associated with the term "task coordination."

We have chosen some typical forms of usage that demonstrate the versatility of the Task Manager tool; as with ordinary paper to-do lists, such different types of usage may be mixed with one another.

Personal to-do list. This is the simplest form of use: tasks in the Task Manager's task list are not shared with other users, but serve as reminders for actions to be taken; these may range from trivial tasks simply represented by their title to specifications of complex projects. Entries in the task list may also simply represent "folders" for a set of tasks.

Brainstorming, conferencing. A task may be thought of as an environment for off-line brainstorming. Prospective contributors are added as participants to the task and are offered a short description as to the purpose of this activity. They may add their ideas as informal notes or documents of any kind. The person responsible acts as a moderator and may form sub-conferences if necessary. A passive observer status is possible.

Meeting preparation. The preparation of a meeting can be made a task: The invitation and other important documents will be distributed to the participants as documents attached to such a "meeting task," the participants can give their feedback via the task conference, and the organization of the meeting, the writing of the minutes, or other follow-up activities can be made subtasks of the meeting task.

Project planning and monitoring. The support of this activity is a natural for the Task Manager: It supports a gradually and dynamically refinable and restructurable structure of tasks and subtasks, the responsibility may be assigned and reassigned, the completion state is reported by the responsible persons as tasks are carried out, and deadlines can be set and monitored.

Project execution. The project work itself can be coordinated, because relevant material and applications may be attached to the respective tasks and subtasks of the project in order to be shared and worked upon by the project members. The somewhat primitive document access and version control mechanisms would have to be complemented for more sophisticated applications by special services.

Repetitive tasks. The Task Manager supports the reuse of task specifications, which may be first edited to fit the current situation and then "pasted" into the list of tasks. Templates for repetitive tasks may be stored privately or copied from organizational databases.

Bulletin board. This again is a very "simple" use of the Task Manager: A task represents a topic described by the title; the bulletins are added as notes or documents to the task and are then available to all participants. Hierarchical structuring of topics is possible via creating subtasks. New participants may be introduced by existing ones. Participants no longer interested may leave the topic in question.

This list is not meant to prove that almost every kind of collaborative (or even noncollaborative) activity may be supported by the Task Manager; of course we are aware of collaborative tasks that would require more specialized support, like the joint authoring of documents. We mean to demonstrate that the Task Manager can be used in various ways, not just for the management of clearly defined tasks.

Dynamic Change of Usage

The various kinds of usage are not necessarily to be kept separate from one another but can dynamically develop from one kind to another. For instance, a task in the personal to-do list ("Should develop a productivity tool for our software production") can be turned into a brainstorming item by adding a few more participants and an initial note about the goal and conditions of the activity. Documents attached to such a personal to-do list item, such as a draft proposal ("A Distributed Task Manager for Software Production"), are automatically distributed to the participants of the brainstorming task and may then be commented on; new versions may be created or substantial comments added as new document resources.

Eventually, a project plan could develop from this activity; some persons, having participated in the planning activity, drop out of the task, and new participants are introduced who have special skills needed for the project. A senior manager is added as an observer of the top-level task so as to be informed of the essentials of the project. The project plan is implemented as a task structure of the Task Manager (hierarchical refinement and dependency networks); responsibilities are assigned, and initial requirements and design documents are attached to the task. Ongoing work is reported via task notes or by state change (e.g., from "not finished" to "finished").

Deadlines are monitored by the system; necessary changes may be negotiated via the task conference, and implemented during execution without any difficulties. Responsibilities for arbitrary subtasks may be reassigned. A history of the project execution is kept and can be reviewed. Up-to-date documentation is available at any given time. Hopefully, the project will be terminated successfully some two years later. If not, there is still a documentation of the case from which to learn for future occasions.

The example we have just given describes the case of a simple personal task developing into a structured multiperson project in a streamlined fashion. It is easy to show that there are other possibilities of interchange between different types of usage as well; for example, a clearly defined collaborative task with a deadline and all turns out to be not that clearly understood during the initial phase of task performance; the participants may need to step back and "misuse" an exchange of informal notes within the task to brainstorm about alternative and more suitable ways of organizing the particular task. In the extreme, all participants could "leave" the task and the person responsible might end up all alone with a now personal task and the necessity to reorganize.

Synchronous and Collocated Usage

Within the well-known two-by-two matrix of CSCW (computer-supported cooperative work) systems introduced by R. Johansen, which distinguishes systems along the dimensions of temporal and spatial distribution, the Task Manager clearly falls into the asynchronous, geographically distributed category: The Task Manager does not require that its users sit in front of their workstations at the same time nor in the same room. On the other hand, it does not forbid this—that is, the Task Manager may also be used in the collocated or synchronous case.

For instance, a user may realize that another user is actively manipulating tasks around the same time and could react by sending a note or even phone the other participant and discuss rearranging or rescheduling a group of subtasks, while both would look at the task list and do some editing. They could also sit in front of the same display and work together on some tasks; the results of such a session would be automatically distributed to each of them (and any other participants).

Another example is that of a project meeting where the task structure is discussed by the participants equipped with the Task Manager on mobile computers. So, although mainly meant for work situations distributed in time and space, the Task Manager may also be used synchronously or in collocated situations.

Scalability of Use

The Task Manager may not only be used by a limited number of users over a local network. Special attention has been paid during its development to its scalability with regard to the number of users, the number of tasks, and the dimensions of geographical distribution. The Task Manager is not limited in this respect other than by the availability of computer storage and store-and-forward communication facilities. It may be used for the support of large user communities over large geographical distances without restrictions or deterioration of functionality.

Implementation of the Task Manager

Starting with the conception of the coordination model, the Task Manager was developed over a period of two years, with a first prototype after about 15 months. The subsequent phase of evolutionary development included stabilization of the prototype; evaluation in simulated work situations by potential users, which brought valuable feedback to the developers; and essential enhancements of functionality and user interface toward a workable system.

Having had an early working prototype paid off in terms of quality of the present system. Now that the Task Manager is rather stable, it is in use within the developers' group, among other things for project management and the development of the next version of the system itself. We plan to gradually extend this system usage to different types of settings, such as with the partners of our present European project.

Figure 4.2 sketches the software architecture. It shows two domains linked by standard X.400 store-and-forward message transfer—however, the number of domains is not restricted. The concept of domains reflects the different speed of LAN (local-area network) or WAN (wide-area network) communi-

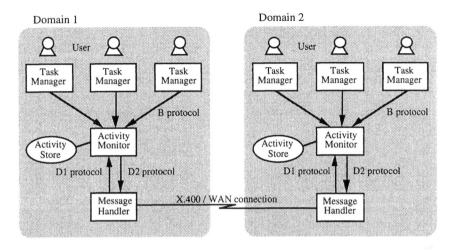

FIG. 4.2 Software architecture of the Task Manager.

cation. Within a domain, a client–server approach is used to update a shared task structure, while X.400 messages are used to distribute changes to other domains. Every user of the system is located in a specific domain and uses his or her own instance of the Task Manager. This tool manages user-specific views of the user's task list. Using remote procedure calls, the Task Manager talks to the Activity Monitor to get up-to-date data and to request changes to the domain-wide shared Activity Store. In each domain, there is one instance of the Activity Monitor that serves multiple Task Manager instances.

When a user modifies a specific task or the task structure, the Task Manager sends an operation request to the Activity Monitor. The monitor executes the operation in the Activity Store and broadcasts the request to the remote domains involved via the Message Handler. When an operation request arrives from a remote domain, the Message Handler passes it over to the Activity Monitor, which then updates the corresponding task objects in the Activity Store accordingly.

The task objects are actually replicated in different domains. This is needed because user operations on tasks should be immediately reflected in local changes. A user will not wait for operation requests or results transported over WAN connections, but requests a fast response—even if the response is preliminary in some cases, as discussed later. Strictly speaking, tasks have to be replicated not only in those domains where participants of the task itself are located but also in domains where participants of supertasks can observe a task (a task can be subtask of more than one supertask). Therefore, the replication of task objects depends on task participants and the hierarchical task structure, which both may change dynamically. Task replication is completely transparent at the user interface. The essential purpose of the Activity Monitors is exactly the distributed control of the replication and modification of task objects, in spite of WAN connection breakdowns between Activity Monitors and lost or duplicated messages on WAN connections.

Obviously, a concurrency problem has to be solved for this architecture: Operation requests arrive asynchronously at the Activity Monitors from local and remote domains, whereas all users in their different domains should eventually have a compatible view of the task structure (which need not necessarily be the same view). As a solution, we have designed the D protocols between the Activity Monitor and the Message Handler in such a way that the causal order of operations on task objects is preserved.[2] As a main feature, our protocols enable the Activity Monitor to detect concurrent updates of task attributes, and to resolve conflicting assignments by assuming a linear order on those events.

Of course, all Activity Monitors must reach the same belief about the linear order eventually. The effect of this strategy for users is that sometimes local changes become "overwritten" by changes due to remote participants. Because all changes are highlighted at the user interface, a user at least becomes aware of what has happened.

The asynchronous distribution mechanism for tasks is also used for the distribution of notes, documents and services. Lists of notes, document references, and service access points are attached to tasks and technically managed as task attributes. In order to attach or to change a document, the document or new version is copied with a new filename to a host in the local domain, and the pair "host/filename" is distributed. When accessed remotely, the documents are sent by file transfer protocol (ftp) (a cache can be used on the remote site to avoid multiple remote copies).

Although the distribution of new versions of documents admits only an asynchronous, coarse-grain update of documents, the usage of services allows for

---

[2] Our approach is similar to the *cbcast* method within process groups in the ISIS system (Birman et al., 1991). But in contrast to the ISIS system, we only need causal order to deal with changes of a tasks' participant set, and such changes are not atomic (membership changes of process groups in ISIS are atomic).

distributed editing and synchronous communication. Task Manager and the service use a simple protocol to agree upon what X server the service is to be used on when it eventually contacts the user. All services integrated in our system must obey this start-up protocol.

The concept of documents and services renders the system "open-ended." We support documents of any kind as long as their associated tools, like word processors or graphical editors, run on the underlying UNIX operating system.

The Activity Monitor is implemented as an ISO-ROS service (Rose et al., 1991) supporting two access protocols, B and D_1. The Message Handler offers another ROS service: Via the D_2 protocol the Activity Monitor propagates its data across domains. The protocols are defined in ASN.1 (Steedman, 1990). All clients and services of the system are implemented in C++ using ISODE (Rose et al., 1991).

Assessment of Usability and Usefulness

First, we compare the Task Manager with our initial design goals.

Flexibility: We have already shown that the Task Manager is a versatile tool that may be useful in a variety of work situations. Tasks may be performed by just one user or shared by any number of users in the same way, a shared task may be structured to various degrees of refinement and dependency, and all attributes, including the participating users, may be changed dynamically. The only restriction to flexibility perceived so far is the rather egalitarian model of task sharing: All participants share the same view. Extensions to the tool might be needed in order to adapt to rather different organizational cultures than we originally had in mind.

Interoperability: The concept of attaching documents and services to tasks opens up various ways in which existing tools or applications to be created may interoperate with the Task Manager. First, arbitrary documents attached to tasks may be opened and worked on without leaving the task management context. Second, services offer specific cooperation support not covered by the Task Manager within the task management context.

Self-organizability: Users of the Task Manager may organize their collaboration as they see fit, with the only prerequisite being the installation of the tool in the user community. There is the possibility of using and adapting task templates from an organizational database, or reusing old task specifications.

Simplicity, genericity: The Task Manager in its present form is basically a very simple tool, and is usually very quickly understood by novice users as a distributed to-do list. It may be applied to any kind of collaborative activity, because it has not been tuned to any specific application domain.

Early designs and a first prototype of the Task Manager were evaluated in user workshops conducted in the spirit of the Scandinavian approach of user-centered design (Kyng & Greenbaum, 1991). Our user organization was a company that manages a very large technical project, and our users at the workshops were managers, engineers, and support personnel; details on the organization may be found in Grønbæk et al. (1992).

The main goal of involving prospective users was to find out whether the Task Manager addressed the problems of distributed work management found in the user organization, and what might improve its design. The prospective users regarded the Task Manager as useful for coordination within their organization, but thought overview and efficiency of the tool should be improved. This criticism has led to considerable enhancements of the Task Manager.

Finally, we found it interesting to compare the Task Manager with the checklist for a successful CSCW tool that recent analyses of cooperative work—computer-supported or not—came up with. Robinson (1993) claimed that a tool should have a clear functionality to do a job and in addition should support the following features, which we think are also present in the Task Manager:

Peripheral awareness is offered by the task list interface, which at a glance notifies of events in shared work the user should attend to.

Implicit communication is supported via sharing documents and services attached to a task.

"Double level language," that is, the complementary and mutually supportive use of implicit and explicit communication, is provided for by the conference of informal notes attached to a task.

Overview of distributed work is one of the main features of the shared to-do list.

Multifunctionality could be claimed as already demonstrated, but has yet to be proven in real use.

Our first assessments have shown that distributed to-do lists can contribute to the management of distributed work. This has encouraged us to proceed with the further development of the Task Manager.

Conclusion

There is a still growing need for tools to manage distributed work. The coordination models and systems developed over the last years suffered from the rigidity of predefined procedures and imposed structures, and from the isolation from informal communication and other forms of computer support.

With the Task Manager we have presented a simple, powerful, and sufficiently generic tool for the management of distributed work that addresses these problems by its flexibility to adapt to a broad variety of collaborative work situations and by the possibilities it offers to interoperate with other computer support.

4.2 COMPUTATIONAL DIALECTICS

Thomas F. Gordon

The Rationality Crisis

The central task in practical problem solving is to identify and choose among alternative courses of action. A couple must decide which car to buy. The designers of the Dylan programming language had to decide whether its syntax should be more like LISP or Algol. Volkswagen must decide whether to manufacture the new "Beetle" shown at a recent international automobile show. The editors of the *General Anzeiger* had to decide whether to put the story about the burning of a housing complex for asylum seekers on the front page or bury it near the back.The Social Democratic Party had to decide whether or not to include an Autobahn speed limit in their platform for the upcoming election.The German parliament had to choose between Bonn and Berlin to be the capital city of the reunited German state. The United Nations and NATO must decide whether or not to use military force in Bosnia.

The main purpose and promise of computers and information technology is to improve the procedures for making choices of this kind in industry, government, and other kinds of organizations and groups. The improvement may be in effectiveness, efficiency, or, when the conflicting interests of multiple parties are involved, fairness.

The different subfields of computer science contribute to this abstract goal in complementary ways. When there is perfect information about a problem, an efficient algorithm or theorem prover may be used to compute or search for a solution. Large databases make a wealth of relevant information readily available. Knowledge-based systems are useful for tasks where there is sufficient consensus about the knowledge required, and the costs of knowledge acquisition and maintenance can be amortized over the expected life time of the system. High-capacity networks and hypermedia technology are making it cheaper and easier to disseminate and access all kinds of information, including text, sound, color graphics, and video. So-called "virtual reality" systems and other kinds of computer simulation make it possible to explore and vividly imagine the likely effects of alternative courses of action. Even applications as banal as word processing, spreadsheets, and electronic mail flourish in the end because of their role in the processing and distribution of information to be used in making decisions.

As useful as these technologies have been shown to be, none of them squarely confronts the problem of supporting effective, fair, and rational decision-making procedures under the conditions which usually prevail. Either they only deal with a part of the problem, such as providing access to relevant information, or they restrict their attention to special problem-solving contexts where certain simplifying assumptions, such as perfect information, can be made.

Under what conditions must decisions usually be made? Here are some of the more salient factors:

1. There is both not enough and too much information. For some parts of the problem, relevant information that would be useful for making a decision will be missing. For other parts, there will be more information than the persons responsible for making the decision will have time to even retrieve, let alone comprehend.

2. The resources that can be applied to finding a solution are limited. Time, in particular, may be "of the essence": A solution must be found before the issue becomes moot.

3. The expected value of the known alternative decisions is not high enough to make it cost-effective to invest substantial resources in implementing a program, knowledge base, or other kind of elaborate computer model to use in helping make the decision.

4. However much information is available, opinions differ about its truth, relevance, or value for deciding the issue.

5. Arguments can and will be made pro and contra each alternative solution.

6. Reasoning is defeasible. Whatever choice seems best at the moment, further information can cause some other alternative to appear preferable.

7. Factual knowledge about how the world functions and its current state is not sufficient for making a decision. Value judgments about ethical, political, legal, and aesthetic factors must not only also be taken into consideration, but are the critical issues requiring the most attention.

8. Several persons have a role to play in making the decision and will be affected by it. Conflicts of interest are inevitable; support for negotia-

tion and other procedures for achieving consensus and compromise are required.

9. Finally, the persons responsible for making the decision are not proficient in mathematics, logic, or any other formal methods for solving problems.

Again, this is not a worst-case characterization, but rather a fair and realistic description of the conditions under which decisions must usually be made. Increasing awareness and acceptance of this fact, among both the general public and experts in fields such as philosophy, jurisprudence, and mathematics, has led many people to cast doubt on the whole enterprise of rationality.

Computer science is built upon a conceptualization of rationality coming increasingly under fire. Preserving a proper role and justification for information technology will depend critically on developing the theory, methods, and applications for assisting individuals and groups to make effective and fair decisions under ordinary circumstances.

On the theoretical front, computer science desperately needs to intensify its dialogue with the humanities, including philosophy, law, history, literature and the arts. Effectiveness and fairness are normative concepts. The natural and engineering sciences provide us models of how the world functions and technology for changing the world in sometimes dramatic ways, but they address only the easier half of the general problem of making rational decisions. Knowing what can be done tells us nothing about what should be done. It is the humanities that provides standards and methods for evaluative judgment.

Regarding methods, the metaphor of an assisting computer system, the guiding idea of the AC research program at GMD, is a useful starting point. The mediator, moderator, or arbitrator of a discussion, debate, brainstorming session, or bargaining meeting is a kind of assistant. He or she is not a principal participant in the discussion, but rather has an ancillary function, such as helping to assure that the speakers abide by the rules of the procedure. A mediator has little or no authority. It is neither cop nor judge. The function of a mediating computer system is not to automatically enforce some formal, and therefore rigid, set of procedural rules for resolving conflicts and deciding issues, but rather to advise the participants about the rules and provide other information about the state of the proceeding.

As for applications of this idea, several mediating systems for coordinating the activities of a group have been designed and implemented during the course of the AC program in the Computer-Supported Cooperative Work (CSCW) research division of our institute (Kreifelts et al., 1991, 1993). These systems help groups with such tasks as scheduling appointments and meetings, creating and monitoring plans, and guiding the flow of forms through an organization.

Zeno will be a mediating system for assisting the more generic task of discovering and choosing among alternative courses of action. The Zeno system will be able to moderate a discussion or debate about any topic between ordinary persons with no particular technical skills in computer science or logic. Our ambition is to develop a practical system for supporting decision making in groups under ordinary circumstances.

There is a trade-off between ease of use and functionality. Supporting deep reasoning requires complex formal logics. Ordinary users cannot be expected to express their positions in formal languages of any kind, and the state of the art of natural language processing has not yet reached the point where the translation to and from a suitable logic can be automated. Finding a good trade-off between ease of use and expressiveness that does not require natural language processing is one of the main problems to be addressed by Zeno. We call our current approach lazy formalization. The idea is that the participants in a discussion are free to choose the level of formalization they deem appropriate. In fact, a speaker may use any means of expression desired, formal or informal, textual, graphical, or multimedia. The discussion begins using a logic that is so simple that it can be hidden completely behind an intuitive user-interface.

To give a better idea of the kind of system we have in mind, the next section describes Zeno's current design, from the user's perspective. The following section is more theoretical; it discusses a proposal for a new field of computer science research, to be called *computational dialectics*, whose subject matter is computational models of norms for rational discourse. This field is founded on a conception of rationality that, we claim, can withstand the criticism and concern of the skeptics. The final section discusses related work.

A Tour of the Zeno System

The Zeno system will be configurable for different kinds of deliberations about some topic, such as brainstorming sessions, council or board meetings, contract negotiations, design team discussions, and lawsuits. There will be two interfaces, one to configure the system and another for using a particular configuration to mediate a proceeding. The first interface can be viewed as a high-level programming language for implementing mediating systems. Compiling a program in this language generates a mediating system for a particular type of proceeding. We have more to say about this "programmer's interface" later. Let us first take a look at the interface to be used by the persons taking part in a discussion.

Figure 4.3 shows a mock-up of a Motif version of the main window of the Zeno application. It appears to be a cross between an electronic mail program and a hypermedia browser, and indeed it has characteristics of both.

The "File" menu includes the usual commands for such things as opening, closing and printing documents. A Zeno document contains references to all the messages registered with the mediator for a particular proceeding or task. To open a document, the user must first log in to the mediator's machine on the network, providing a name and password. The rights of the user to view or send some message may depend on such factors as the type of the proceeding and the role of the user in this proceeding. Several participants can open and modify the same document simultaneously; as messages are only added during the discussion and never deleted, the usual synchronization problems of databases and multiuser text editors do not appear here. However, the rules of the proceeding may have to specify when each kind of speech act is to be legally effective; for example, at the time it was sent or at the time of receipt by the mediator.

Instead of the usual "Save" there is a "Send" command. A user can modify the network of claims and arguments locally, playing "what-if" games to see the effects of alternative lines of argument, before sending a contribution back to the moderator. There will be unlimited undo and redo commands as well as a "Revert" command to facilitate this kind of private contemplation.

There will also be a "Save As" command for saving a local copy of the document and for exporting it to other file formats. Of particular interest would be the possibility of exporting an outline of the discussion, or selected parts of it, in the native formats of various word processing, "idea processing," and "presentation" applications. This would be quite useful for writing such things as the "minutes" of the discussion or the justification of a decision.

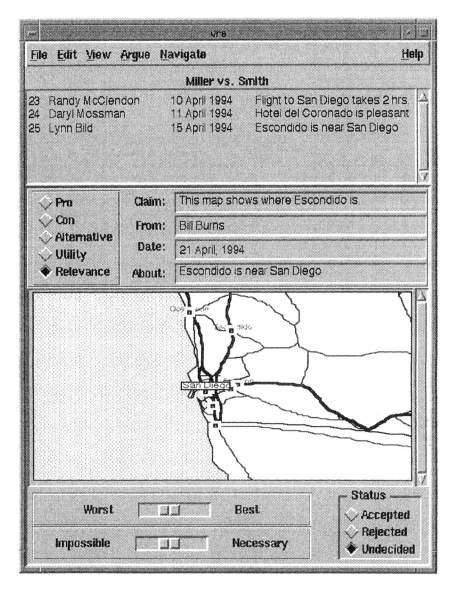

FIG. 4.3. Main window of the Zeno application.

Just below the menu bar, in the center of the display, is the title of the proceeding, in this case "Miller vs. Smith," suggesting this may be some kind of legal discussion. In the area below the title is a scrollable transcript of the messages that have been registered with the mediator. These need not be all, or even most, of the messages that have been exchanged by the participants in the proceeding. It is not intended that the Zeno system be used to replace all other forms of communication within a group. On the contrary, it should be used primarily for those speech acts that are to have some kind of official or binding character.

This brings us to a problem Zeno, like other CSCW applications, must deal with: how to integrate the system with the other applications, to facilitate interoperability, data exchange, and ease-of-use. Presumably most users will already be using some other program for electronic mail. Some may not want to use another system to send messages to a mediated discussion, with yet another set of user-interface conventions and quirks. Although a complete solution to this problem will have to await the widespread use of distributed object-oriented programming environments, an intermediate approach is possible for the time being. First of all it should be possible to cut and paste data between Zeno messages and other applications, at least for the more popular data formats. Second, a simple command language, along the lines of the ones used by network mailing lists, will allow messages to be sent to the mediator using any electronic mail program.

Below the transcript in Fig. 4.3 is some "header" information about the message being displayed, including a short description of the "claim" being made in the message, the name of the sender, the date and perhaps time the message was sent, and a pointer to the message for which this message is a response. The claim can be any unique title for this message. It need not actually be a declarative sentence, although this might be a good practice. In the example, the message claims "This map shows where Escondido is" and is offered to support the claim of the message contributed earlier by Lynn Bild, who claimed that "Escondido is near San Diego."

To the left of the field naming the previous message is a label showing the type of this message, in this case an argument "pro" the claim of the other message. Although the full set of message types will be defined by the designers of a Zeno application, there may be a few "standard" types, such as:

Agree. Used to agree with or concede some other claim.

Disagree. Used to challenge, question or deny the other claim.

Pro. A claim that, if accepted, tends to make the claim of the prior message more likely or probable.

Con. A claim that, if accepted, tends to make the claim of the prior message less likely or probable.

Alternative. Proposes another solution to the problem, or takes another position with respect to the issue.

Utility. Makes an assertion about some effect or consequence of deciding to accept the claim of the prior message. For example, one could claim that a Porsche is a fast car, or that a Volvo is a safe car.

Relevance. Questions whether the prior message really is of the type asserted. For example, suppose Judy claims that it will rain next Saturday and Joe then argues that this is unlikely, because he has planned a picnic for then. Rather than arguing about whether or not he has in fact made such plans, Judy might prefer to question the relevance of his plans to her prediction.

Refinement. Registers a claim that only becomes an issue if there is a decision to accept the claim of the other message. For example, if it has been decided to buy a Chrysler, this kind of message can be used to propose buying a particular model, such as a Voyager.

Comment. Can be used as a "catch-all" message type when none of the other types available seem appropriate, or when the speaker wants to avoid the formal consequences of some other speech act. In some applications, it may also be permitted to send comments, and perhaps other types of messages, anonymously.

To compose a message, the user selects its type from the "Argue" menu. Another name may be preferable for some applications, so this will be configurable in the Zeno programming environment.

It might be objected that users will not want to take the trouble to label the type of their messages, or would prefer to remain vague or ambiguous about the intended pragmatic effect of some speech act. For example, in a message to the boss criticizing his or her plan to manufacture horseshoes instead of tires, one might prefer to tactfully couch the warning in the language of admiration and support. A large, bold label of "criticism" might be counterproductive.

This is admittedly a problem, but not an insurmountable one. One can use the innocuous comment label in such cases. Also, as the body of the message is subject to no formal restrictions, the user is encouraged to apply rhetorical skills, to the best of his or her ability, here.

However, there is a better response to this objection. Performative speech acts are often effective only if they have the proper form, regardless of their intended meaning. In business and government, one often has to say the magic words. There are sound reasons for this formality. The interests of the persons affected by a decision will differ; often they are diametrically opposed. The buyer of some product or service would like the price to be low; the seller would like it to be high. Whenever it is in the interest of one party to speak vaguely, so as to hedge bets by delaying the determination of the message's performative effect as long as possible, there is probably another party to the transaction with exactly the opposite interest, to have the matter clarified definitely as soon as possible. Consider a letter offering to buy some product from a mail order distributor. The seller would like the assurance that it is indeed a binding offer before sending the goods. The buyer would prefer this question to be decided after the goods have arrived, to be able to inspect them before deciding whether or not to pay. The buyer would like to be able to back out of the deal by arguing that the "offer" was really only an "inquiry." Formal procedures and "bright-line" criteria for categorizing speech acts provide the means to fairly allocate risks and responsibility in such situations. Clear conventions also dramatically reduce the cost of doing business, by avoiding lengthy and expensive conflict resolution procedures, such as lawsuits.

There is no universally optimal degree of formality, suitable for all kinds of group decision-making contexts. In particular, CSCW systems that support only informal modes of communication are biased; they cater to the special interests of only some of the persons affected by the decisions made using the system. The aim in Zeno is to remain neutral by providing a configurable environment supporting a wide range of formality. Design choices about this and other aspects of the procedures for making decisions in a group or organization should be made by representatives of the various interests groups affected, through some fair political process.

Once a message has been sent, the rights and obligations of the other participants will change, depending on the rules of the proceeding. For example, in a negotiation, an "offer" message may give some other participant the right to "accept" within 30 days. Or the posting of an issue may require position statements to be made within six months, before the issue comes up for vote. One of the main responsibilities of the mediator is to maintain a calendar and agenda of such tasks. There are commands for displaying these documents in the "View" menu. The calendar shows the schedule of dates and times for various activities. One possible service of the mediator would be to remind users of deadlines. The agenda is a prioritized list of issues to be resolved, where the criteria used to prioritize issues or tasks will depend on the application.

Returning to Fig. 4.3, below the header information is a scrollable pane for the body of the message. In the example, this is a color map of San Diego. Again, there are no restrictions, in principle, on the kinds of data that may be included in messages. From the perspective of Zeno's formal logic, each message is a proposition. As always in formal logic, the intended meaning or interpretation of the proposition is ignored when deriving consequences and other kinds of formal properties. However, the persons participating in the discussion will of course be quite interested in the meaning of a message, which will presumably play a dominant role in their contemplations about how best to respond.

Propositions in Zeno are situated; they are contextually embedded in a discussion between persons taking place in time. A proposition does not hang in the air, but is stated by a particular person at a particular time. Except of course for the opening proposition, every statement is made in response to some other claim made in the course of a discussion. One must be careful when carrying over arguments and claims made in one context to some other context in this or another discussion. Syntactically equivalent claims in different branches of a discussion are not presumed to be identical.

A message may also be a compound document, consisting of a combination of graphics, text, and other objects, including hypertext links to other messages and documents. Unlike the message types discussed earlier, these hypertext links have no particular semantics for the logic of the Zeno system. They may be used in any way a user sees fit and help to reduce the "rigidity" of Zeno's formal logic.

While we are on the topic of hypertext links: As every message except the first is a response to some other message, they form a tree structure. The "Navigate" menu includes the usual commands, familiar from hypertext systems, for browsing this tree. For example, the "Top" command takes the user to the first message of the proceeding; the "Up" command moves to message responded to by the current message; and the "Next" and "Previous" commands cycle through the other responses at the same level. To move to a lower level, there are submenus for each type of response, such as "Pro," "Con," and "Relevance."

To perhaps belabor the point, this graphical interface provides an intuitive way to express the elements required by a formal logic (propositions and various kinds of relations between them) without requiring the use of some formal syntax.

At the bottom of the window in Fig. 4.3 is the final pane to be discussed; it displays information about the current status of the claim. On the left-hand side there are two sliders, showing the logical status of the claim along two dimensions. The first dimension concerns the *quality* of the position, relative to the other proposed alternatives. Quality is computed using the utility arguments which have been made for each of the alternatives (discussed later). The other dimension concerns the likelihood, probability, or feasibility of the position and is computed using the arguments pro and contra that have been made concerning it. Of course, these sliders can not be manipulated by the user, willy-nilly, to set the value of these parameters. Rather, they are continuously computed by the Zeno system, using a combination of theorem proving and constraint satisfaction techniques.

To the right of these two sliders is a group of buttons showing whether the claim has been accepted, rejected, or is yet to be decided. The procedure for making this decision will depend on the rules of the particular type of proceeding. Common methods include randomly selecting some alternative, using the best possible alternative computed by Zeno, voting, or granting the responsible manager or authority discretion to decide as he or she sees fit. Notice that the quality and probability measures computed by Zeno have only an advisory character; they may be taken into consideration by the persons responsible for making the decision, but need not determine it. This flexibility is perfectly reasonable. After all, the system is founded on the premise that reasoning is defeasible. The responsible person may have information that for various legitimate reasons he or she is unwilling to divulge to the group and that tips the scales in favor of some other alternative. Or the individual may simply prefer to follow personal intuitions.

Associated with each claim are three other documents: (a) a worksheet for making and viewing claims about the relative weight or importance of the arguments pro and contra this claim, (b) another worksheet summarizing the arguments made about the relative utility of this claim and its alternatives, and (c) a document for recording information about the decision. This latter document may include such things as the name and "signature" of the person making the decision, the date the decision was made, an explanation or justification of the choice, or a tally of the votes for and against each alternative, as appropriate.

The worksheet for presenting utility arguments is shared by all alternative positions for some issue. It has two parts. The first part is a list of utility claims that have been decided to apply to each alternative position. For example, when discussing which car to buy, the following utility claims may have been accepted:

> *BMW 520i:* good chassis, good styling, good interior, good safety, fair fuel economy.
>
> *Mazda Xedos:* good chassis, fair styling, fair interior, good safety, fair fuel economy.
>
> *Opel Omega:* good chassis, good styling, good interior, good safety, fair fuel economy.

These claims could be displayed in a table, but this will not generally be the case. A utility claim may be an arbitrary proposition about some effect of choosing the alternative; Factors or dimensions, along which the alternatives would be ranked, need not be first systematically identified.

The second part of the utility worksheet is a list of "constraints" about the relative values of these utility claims. In the car buying example, the following evaluation constraints may have been accepted:

- Good interior › fair interior.
- Good safety › good interior.
- Good safety › fair fuel economy.
- Fair interior + fair fuel economy › good safety.

The main purpose of these constraints is to provide an easy, qualitative way to express and argue about preferences and value judgments. It is not necessary here to devise factors and utility functions, let alone assign numeric values to particular properties. Given this qualitative information, constraint satisfaction techniques can be used to rank the quality of the alternative solutions.

In the graphical user interface, there will be some intuitive and quick way to go to the message in which it was decided to accept some claim shown on this worksheet. For example, to find out why it was decided to believe that an Opel Omega has a good chassis, one might be able to just double click on that property on the worksheet to begin browsing any arguments there may have been about the quality of the chassis.This applies to the evaluation constraints as well, which are debatable just like other claims.

It remains to discuss the "programmer's" interface for configuring Zeno for a particular kind of discussion or proceeding. Some of these discussions will be primarily cooperative, and others will be more adversarial. Other factors to consider when drafting the rules of the procedure include its goal and purpose, the types of speech acts required, and the roles of the participants. The rules of procedure will specify just what speech acts are permitted, obligatory, or forbidden in each situation, and at what time, where a situation consists of the messages that have already been registered with the mediator.

A configuration also needs to specify what the mediator should do in the case of a violation, or attempted violation, of the rules. However complex the rules, situations are likely to occur that were not anticipated. One way to avoid rigidity when configuring Zeno is to use the legal system as a model. Unlike formal systems, legal rules are not self-applying. Persons must interpret, and reinterpret, the rules in the context of their current situation. In the worst case, a lawsuit may be necessary to resolve disagreement about the meaning of the rules. In a Zeno application, this strategy could be realized by having the mediator send a private warning to the persons affected, who would then have the option of negotiating a "settlement" or initiating some quasi-legal procedure for resolving the conflict.

The language for defining these rules has yet to be designed. It is still unclear whether a simple and convenient graphical user interface will be possible for configuring Zeno. Arguably, it is not quite so important for this interface to be easy for lay persons to use, as some small number of configurations will be adequate for most purposes. Experts could be hired to help design and implement a custom configuration. On the other hand, it is critically important that every user be able to understand the rules of the proceeding, so as to be able to effectively participate and decide whether or not others are "playing by the rules."

Computational Dialectics

Zeno is but one project in the field we call computational dialectics. The subject matter of this field is the design and implementation of computer systems that mediate and regulate the flow of messages between agents in distributed systems, so as to facilitate the recognition and achievement of common goals in a rational, effective, and fair way.

The term "agents" here is intentionally abstract. An agent may be a person or organization, or some computational entity, such as a process, task, or object, in the sense of object-oriented programming. In a complex, distributed system consisting of multiple agents working together, some of the agents will be natural persons or organizations, and others will be artificial agents implemented by programs executing on one or more computers.

The field of computational dialectics has its analytic, empirical, and normative aspects. The analytical task is to develop models of the structure of discourse and communication tuned to the task of group problem solving and decision-making. This distinguishes the models of dialectical processes from those designed for understanding natural language. The analytical task, as usual, consists in identifying, categorizing, and analyzing the formal properties of these models along various dimensions. The empirical aspect involves developing and testing theories of how, in fact, groups of agents use language to make decisions. Finally, the normative aspect of the field is concerned with drafting and justifying principles and norms for regulating communication and decision making in groups, where individual agents may have incompatible beliefs about the world and competing interests.

To be sure, much prior work has been done in this area, if not under this label. It is our hope and goal to bring together researchers who have been working implicitly on this subject in the fringe of other parts of computer science, including distributed systems, distributed artificial intelligence, nonmonotonic logic, case-based reasoning, machine learning, conflict resolution in concurrent engineering, artificial intelligence and law, issue-based information systems, and computer-supported cooperative work. Presumably, research in computational dialectics would be more productive if the people interested in this subject would begin to form a research community. As a first step in this direction, we organized, together with Ronald Loui, a workshop on computational dialectics for the Twelfth National Conference on Artificial Intelligence (AAAI-94). Additional work is needed to reach people outside the AI community.

The thesis of the Zeno project, which represents only one position in the field of computational dialectics, is that rationality can best be understood as a theory construction process regulated by discourse norms. The dominant conception of logic in analytical philosophy is limited to the study of the notions of consequence and contradiction given some set of premises. It says nothing about how the premises are or should be constructed. However, by viewing rational discourse as a process of theory construction, a strong connection to logic is preserved. Our aim is to complement logic with norms regulating the pragmatic aspects of constructing and using theories.

Related Work

Prior work of the CSCW group at GMD on coordination systems was mentioned in the introduction (Kreifelts et al. 1991, 1993). Again, whereas these systems support the scheduling of meetings, the monitoring of tasks and activities, and the flow of forms through an organization, Zeno mediates a discussion about the pros and cons of alternative solutions to a problem.

Several other hypertext systems have been constructed for organizing and browsing arguments, based on either the Issue-Based Information Systems (IBIS) model (Conklin & Begeman, 1988) or Toulmin's model of argument structure (Marshall, 1989; Toulmin, 1958; Schuler & Smith, 1990). The argument structure designed for Zeno is a synthesis of ideas from these systems. Unlike Zeno and the Pleadings Game, discussed next, these other hypertext systems do not use logical dependencies to constrain or facilitate the further development of the discussion. The goal in Zeno is to achieve the simplicity and ease of use of IBIS without sacrificing a solid, logical foundation, by drawing on the results of argumentation systems for nonmonotonic logic (Geffner & Pearl, 1992; Pollock, 1988; Simari & Loui, 1992). With the exception of the Pleadings Game, none of these other systems distinguish the roles or interests of the persons involved in the discussion, so the idea of regulating argumentation using discourse norms does not appear.

The Pleadings Game (Gordon, 1993, 1995) is a computational model of a mediator for a particular kind of legal proceeding, the pleading phase of a civil case. Pleading is a two-party adversarial procedure whose purpose is to identify the issues of the case. The plaintiff has the burden of defending his claim against various kinds of attacks by the defendant.

The Zeno system generalizes the Pleadings Game in a number of ways. Although the Pleadings Game is a particular mediating system for one kind of decision-making procedure, the goal of Zeno is to provide a convenient language for specifying a broad range of mediating systems, for both cooperative and adversarial contexts. Another important difference is that the Pleadings Game model has an entirely theoretical purpose, to demonstrate how judicial discretion can be fairly and rationally limited by factors other than the literal meaning of legal texts. The purpose of the Zeno system, on the other hand, is to provide a practical tool for implementing systems which mediate actual discussions between persons.

Notwithstanding these differences, both Zeno and the Pleadings Game are based on insights from legal philosophy, especially the normative theories of legal argumentation of H. L. A. Hart (1961) and Robert Alexy (1989).

Hart and Alexy, in turn, both draw heavily on the "speech act" theory of language going back to (late) Wittgenstein. There is an ongoing controversy within CSCW about the suitability of speech-act theory as a basis for computer systems for coordinating human activity in organizations. A recent issue of the Computer–Supported Cooperative Work journal included two articles on this very issue: one by a critic, Lucy Suchman (1994), the other by Terry Winograd (1994), who together with Fernando Flores first introduced the use of speech-act theory to CSCW in the influential "Understanding Computers and Cognition: A New Foundation for Design" (Winograd & Flores, 1986).

Legal philosophy provides another perspective on this issue, which reveals weaknesses in the arguments of both Suchman and Winograd.

Suchman took the position, closely related to Grudin's (1990), that "the adoption of speech-act theory as a foundation for system design, with its emphasis on the encoding of speakers' intentions into explicit categories, carries with it an agenda of discipline and control over organization members' actions." In other words, she claimed this kind of CSCW system furthers the interests of management at the expense of workers. She proposed instead that CSCW systems be designed with "an appreciation for and engagement with the specificity, heterogeneity and practicality of organizational life."

Winograd countered by arguing, in essence, that a certain amount of rigidity and formality is a necessary evil in large organizations: "When people interact face to face on a regular day to day basis, things can be done in a very different way than when an organization is spread over the world, with 10,000 employees and thousands of suppliers." And further, "The use of explicitness makes possible coordination of kinds that could not be effectively carried out without it."

If we identify corporate interests with the interests of management and suppose that these interests conflict with those of employees, then Winograd may be thought to be conceding Suchman's main point here. However, he goes on to argue that coordination systems can be successful only if they are "grounded in the context and experience of those who live in the situation."

To assure this is so, Winograd argues that users should participate in the design of the system.

At first glance, there may not appear to be anything new or interesting about this debate from a legal perspective. Surely it is noncontroversial that changes in the rules of an organization, whether or not brought about by the introduction of new technology, have a political dimension requiring fair procedures for negotiating an acceptable compromise balancing the interests of all concerned.

What does make this debate interesting from a legal point of view is its close relationship to an old debate in legal philosophy about the status of legal rules. In the previous century, German conceptualism (Begriffsjurisprudez) adopted a deductive view of legal reasoning. In modern terms, it sought to apply the axiomatic method to the law. The resolution to any conceivable legal dispute was contained, implicitly, in the axioms, waiting to be discovered by a process of deduction. This view depends critically on the "correspondence theory of truth," which underestimates the difficulty of deciding whether the concrete facts of a case should be subsumed under the general terms used in a statute. This is where Hart comes in. Hart recognized that the meaning of laws cannot and should not be fixed at the time of their enactment by a legislature. Rather, the meaning of the law must be continuously reinterpreted and reevaluated in the context of deciding specific cases in the courts. Hart noted that the ability of natural language to be imprecise is a feature, not a defect; it allows power to be delegated to the courts to decide issues in the context of concrete cases, when more information is available. This line of reasoning leads to a justification of the division of powers between the legislative and judicial branches of government.

Suchman's main mistake is to conclude that rules framed in terms of general "categories" only serve the interests of a particular class, management. There are at least two problems with this position. The first is that the rights and interests of employees, too, may be protected only by this kind of general rule. The "technology" of the language of laws, rules, and agreements is interest neutral. The second problem is that the moral principle of "universalizability" requires norms to be expressed in terms of general categories, rather than concrete situations. This derives from the notion of equality under the law. The tension between equality and doing justice to the "specificity, heterogeneity and practicality of organizational life" is resolved by interpreting and reinterpreting general rules to decide issues raised by concrete cases.

Is the formal structure of speech acts in Winograd's kind of CSCW system like a system of laws? It should be but is not. The problem is that these formal structures have been used to define and create the space of actions, rather than the space of rights and obligations. They have been used to define what is possible, rather than what is ideal. It is not enough to allow users to participate in the design process. Users, too, are not omniscient; they cannot foresee all the possible consequences of an abstract set of norms, divorced from the concrete facts of particular situations. It should be possible to do what is best, and not merely that which is obligatory given a strict, formal interpretation of the rules.

Zeno is modeled after the legal system. The formal rules of a decision-making procedure are not used to limit the space of possibilities. Users remain free to take responsibility for their own actions. They may, at their own risk, violate the formal rules. The mediating system is neither the long arm of the law, nor of management. Its job is to advise users about their rights and obligations, not to enforce the rules. Procedures will be provided for resolving disputes about the meaning of the rules, analogous to court proceedings.

Conclusion

The central task in practical problem solving is to identify and choose among alterative courses of action. Computer science has failed to provide adequate tools for supporting rational, effective, and fair decision making under the conditions that usually prevail. Especially, computer science has yet to develop models of rational decision making in groups that adequately take into consideration resource limitations or conflicts of interest and opinion. This section provides an informal overview of Zeno, a mediating system for supporting discussion, argumentation and decision making in groups, which explicitly takes these considerations into account. Also, a new subfield of computer science is proposed, computational dialectics, whose subject matter is computational models of norms for rational discourse. Zeno is a contribution to this field, based on the thesis that rationality can best be understood as theory construction regulated by discourse norms.

Acknowledgment

Earlier versions of this section were read by my colleagues Barbara Becker, Gerhard Brewka, Peter Hoschka, and Hans Voss. I thank them all here for their constructive suggestions and encouragement.

4.3 ORGANIZATIONAL KNOWLEDGE ASSISTANT

Wolfgang Prinz

The emerging research area of computer-supported cooperative work (CSCW) has yielded a number of groupware applications in the recent years. Most systems address a specific application area, such as joint editing, video conferencing, workflow management, and so on. Common to most of these systems is that they require information about the organizational context in which they are used. This is particularly required when systems are used in a large organization or for the support of interorganizational cooperation.

Following from this requirement, this chapter presents the design and functionality of the organizational knowledge assistant TOSCA[3], which was developed in the framework of the assisting computer as an assitant for the provision of organizational knowledge. It was designed as a supportive service to applications and as a tool that assists users orienting themselves within their organization.

In this section we describe the requirements for an organizational information system, the object-oriented data model that is used for the information representation, the architecture of the overall system, and the design of the user interface that presents and provides access to the multimedia information. We conclude with a description of how this system assists a task management system (see section 4.1).

Requirements

Cooperation in teams and organizations is embedded in an organizational framework. Thus, the provision of information about the organizational context in which users work helps to choose the right patterns for communication and cooperation. Information must be provided to answer questions such as: Who is responsible for carrying out a specific task? Whom can I ask for help? Furthermore, the system should provide information as to how particular tasks are handled in the organization. What are the organizational rules

[3] The Organizational Information System for CSCW Applications. The work described in this section has been partly supported by the ESPRIT basic research action COMIC.

one has to consider? Whom do I have to ask first? Which document type do I have to use? All this information belongs to the knowledge that is normally not or only very implicitly provided by CSCW applications, although it plays a significant role in cooperation. TOSCA provides this information to users and applications.

Comprehensive cooperation support benefits when the resources of cooperation and for the cooperation support, such as documents, calendars, and structured message types (Pankoke-Babatz, 1989), can be integrated with the context in which they are used. TOSCA is designed to be more than a storage server for this information. It allows the association of this information to its organizational context, by linking it to the projects, departments, and so on where it is used or to the people who use it.

The dynamic nature of organization makes it impossible to develop a single representation that fits all considerable organizations. Thus we aimed to develop a tailorable model that allows an adaptation of our system to various organizational settings. The object-oriented approach that was chosen proved useful for that. Together with the provision of an object modeling tool, TOSCA provides visibility of the concepts and allows users and groups to tailor the object model to their specific need.

With its potentially worldwide distribution, its methods for distributed management, and its standardized service interface, the X.500 directory (X.500, 1993) fulfills the requirements for a distributed address directory and scalability. Thus it is worth considering the X.500 Directory as a basis for an enterprise information system (Hennessy et al., 1992). However, shortcomings arise when the directory is applied to a more detailed modelling and administration of organizational information in office environments (Prinz, 1990). Major problems deal with the representation and modelling of organizational relationships and data integrity (Prinz & Pennelli, 1992). Nevertheless, in order to benefit from the existence of X.500 for the provision of external organization information, we found it important that TOSCA integrates access to the X.500 world. Accordingly, the TOSCA object model was designed to provide a mapping of X.500 data objects onto the TOSCA objects at runtime.

Organizational information is of particular importance in large geographically distributed organizations and for the support of interorganizational cooperation (Engelbart, 1990). This raises the aspect of scalability which we see as a crucial issue for the success of CSCW applications. From the administrative viewpoint it must be easily possible to extend the number of users of an application. This requires an underlying distributed service environment, which provides a set of common services needed by cooperation support service. The organizational information service presented here is one fundamental component in such an environment. As a support service it simplifies the introduction and use of new applications into the working environment and this may increase the acceptance of these services (Markus & Connolly, 1990).

In order to provide the best possible representation for various kinds of information, an organization information system must be capable to represent different types of media. Thus, TOSCA allows, besides textual data, the representation of graphics, pictures, photos, audio, and video information.

This section is organized as follows. First we present the model for the representation of organizational information. The design of an organizational browser is described second. Then, the architecture of the system is outlined. The section concludes with a brief description of future plans and a summary.

Representation of Organizational Information

Three different models for the representation of organizational information are distinguished:

- The conceptual organization model,
- The organization object model,
- The meta-object model.

On the conceptual organization model level, organization information is considered to be information about the entities of an organization that determine and describe the working context of users. This includes information about the employees, projects, roles, committees, departments, locations, and so on of an organization. Furthermore, the resources of cooperation such as documents, calendars, and other kinds of commonly used data must be considered. In order to provide helpful information on how to perform tasks in an organization, the system needs to represent guidelines that can be used as resources for the organization of a cooperative activity.

All these discrete bits of information become expressive only when they can be related to each other. Therefore we need ways to describe organizational relationships, such as who is a member or leader of a project, which projects are undertaken by a particular department, who is the projects secretary, who is occupying the role of the technical administrator of a special file-server, who supports which task, or which forms do I need to apply for an organizational procedure? It is also necessary that these relationships can be defined in a dynamic way according to the organizational rules. For example, if a committee consists of the members of the projects of a department, we do not want to list all these people explicitly as would be required in X.500, but we want to express this by an appropriate rule. This reduces redundancy and management overhead and increases consistency when the information is changed, for example, when a new member is added to a project.

One can see that the conceptual organization model can be compared with a semantical net. Accordingly, we distinguish on the meta-object model level between so-called organizational objects and organizational relationships. If we take the data description column of the Zachman framework (Zachman, 1987) for an information system architecture, the organizational objects fall into the category of "business entity" and the relationship objects into the category of "business model."

The definition of each type requires the specification of several properties, which are described in the following two subsections. Then the object model designer is briefly described, which allows the construction of the organization object model based on the meta-object model.

The Organizational Object Type

Organizational object types are used to define a schema for the representation of organizational components. Examples are object types for the representation of employees, project descriptions, roles, committees, message types, rooms, buildings, and so forth. The object model that is used to represent this information in an appropriate way requires the following information for the specification of an organizational object type.

> *Type Name and Super Type.* The name is a unique identifier for the object type. It should be characteristic for the type, that is, it should describe the semantic of the type. The object model provides single inheritance, so that for each type one super type must be identified.

> *Scope.* Assigning a scope to a type is interpreted as an access right on the type, which specifies that this type and its instances can be used only in a special organizational context, such as a project or group. This becomes useful when types are user-defined to avoid a proliferation of types throughout the whole distributed system.

The Corresponding X.500 Object Class Name. For each object type the name of the corresponding X.500 object class[4], if available, must be provided. This information is used to map X.500 entries that have been retrieved from the Directory onto the appropriate object type in the organizational data model.

The Naming Authority (The Naming Concept). Each object has its distinguished name. For the provision of distributed management this name is built using a hierarchical naming schema similar to the one that is used in X.500. One key component of that model is the existence of so-called naming authorities. A naming authority is an instance that is allowed to assign a relative name to an object that is named within its naming domain. The naming domain are those objects that are direct successors of the object in the name tree. Accordingly, naming authorities are inner nodes of the name tree. The name of an object is build by concatenating all relative names of superior objects in the name tree starting at the root. For example, the distinguished name of the institute FIT at GMD is: "DE, GMD, Institute FIT."

This information mainly guarantees a consistent naming in an object system that is cooperatively administered. Combined with appropriate rules it is used to build supportive management tools that automatically place objects at the right place in the naming tree.

The Administrator Specification. A major problem of an organization information system is the administration and maintenance of the system. There will not be a single administrator for the whole information base. People from different organizational domains are responsible for the content of different attributes of an object. This can be expressed using this type property. For each attribute or group of attributes, the responsible administrator can be identified either by name or by the specification of an expression that describes the administrator based on the organizational context of the object. This is, for example, useful when each organizational unit has its own administrator for a particular service.

Default Access Rights. Default access rights for instances of the object type can be specified here. They can include expressions such as "grant read access to all members of the project to which the instance is assigned as a resource," to allow a flexible modeling.

---

[4] The notation X.500 object class corresponds to the notation of an object type in our object model.

The User-Friendly Name Specification. The distinguished name of an object is often not expressive or user-friendly enough for its use in a user interface. Therefore a user-friendly name can be defined for an object type. This name can be built by a combination of attributes of the object as well as by retrieving information from objects that this object is related to.

For example, the user-friendly name of an employee object type can be built by concatenating the values of the attributes given name and surname followed by a colon and the value of the position attribute (e.g., Director) or alternatively by getting the name of the role object that this object is related to with the "occupies role" relation.

A syntax has been defined for the specification of a user friendly name. In the current implementation, only one user-friendly name specification per object type can be defined. It is our intention to allow this to be done on a user basis.

Description. Each specification should contain a description of the object type, describing its main characteristics and usage. In a large distributed system this is useful for both administrators and users who stumble over an object of an unknown type. Very often the pure type schema is not sufficient for an understanding of the type.

Graph Description. To supply a description of the context in which an object is embedded, it is very helpful to provide a graphical view. Because the context is different for each object type, this view cannot be generalized. Therefore, the data model allows the description of a graph layout for each object type. They are very useful for the creation of organizational charts, which provide a graphical representation of the embedding organizational context of the focused object.

Constitutional and Optional Attributes. The object model distinguishes between constitutional and optional attributes of an object type. Constituent attributes are those attributes that are essential for the characterization of an object and that cannot be left open when describing an object type. Therefore a user interface enforces the provision of values for these attributes. Optional attributes are those attributes that provide additional useful information that is not strongly required.

Attributes can be basic data types, but they might also contain picture, audio, or video information (see the subsection on the user interface). Furthermore, attributes can contain expressions that are evaluated on access. These expressions are used to refer to other objects, to express general rules, or to compute a value from other object attributes.

Constitutional and Optional Relationships. Relationships between organizational objects are represented by objects of a special relationship object type. Like attributes, relationships can be either constituent or optional. Accordingly, for each object type a list of constituent and optional relationships must be supplied.

Methods. Finally, for each type a set of methods can be defined. Because TOSCA has been developed in C++, the definition of a new method requires a recompilation. To supply a means for rapid prototyping, the computable attributes that have already been mentioned are provided. They allow the description of query scripts.

The Organizational Relationship Type

Apart from the modeling of organizational entities, the representation of the various relationships between these entities plays an important role for a comprehensive modeling of organizations. Several alternatives have been discussed for an appropriate modeling of relationships in an object oriented systems (Moser et al., 1991, Snyder & Lynch, 1991). For our system the approach to model relationships by particular objects has been chosen. Accordingly for each real-world relationship a corresponding object type must be defined. The organizational relation object type has the following characteristics:

Type Name and Super Type. Same as for the organizational object type.

Corresponding X.500 Attribute Name. For each relation object type the name of the corresponding X.500 attribute type, if available, must be provided. This information is used to map X.500 attributes that have a distinguished name syntax, that is, that point to another entry, onto the appropriate relation object type in the organizational data model and vice versa.

The Relation Identifiers. A relation object describes a relationship between two real-world entities, which are themselves represented as objects. Depending from which entity the relationship is viewed, it needs to be denoted differently. For example, a project membership relation between a project and an employee object is called "has members" from the project view, but "is a member of" from the employees view. These identifiers for a relation are defined in the source identifiers and destination identifiers.

Sorting Criteria. Sorting criteria for the output of the related objects can be specified. A membership relation, for example, which relates employees to a project object, can specify that all employee objects should be sorted by the surname on retrieval. For other relations it might relevant to group objects by types (tools of a room), or by a location attribute (member organizations of an international project). This type characteristic is a default behavior for all instances of a relation type. User interfaces might specify other sorting criteria when they access relations.

Value Set Attributes. By default, each relationship object contains two attributes (source and destination) that contain the description of the related organizational objects. These objects can be described by naming them or by expressions that allow for a dynamic description of the objects that are involved in a relationship. These expressions allow the description of organizational rules such as, voting members of this committee are the project leaders of all projects of the department. Also, they can be used to reduce redundancy by describing rules such as, employees of this institute are the members of all projects of this institute. Furthermore it is possible to define user-dependent rules. This is needed, for example, when the person who is responsible for a task depends on the user's membership in a project. In this case the actual user identity is needed to answer a request. This is useful to support a role-based addressing of messages.

Default Access Rights. Similar to the access rights on organizational objects, default access and modification rights can be defined for relationships.

Attributes. If required, additional attributes for a further description or context-dependent restriction of a relation can be defined. That is used, for example, to describe temporal relationships, that is, relationships that exist only for a certain time interval (e.g., project or committee membership).

Creation and Extension of the Organization Object Model

Based on this meta-object model a set of object types has been defined as the organization object model. These can be regarded as the basic toolbox which can be adopted and extended for specific needs.

For that purpose a window-based object model designer has been developed that supports the creation and extension of the organization object model according to the specification of the meta-object model already described. Thus, the creation of the organization object model based on the meta-object model is supported, as well as the transformation to the implementation level, which is in our case C++ and an object-oriented database (Ontos). The design tool is realized as an application of the organizational information service using the schema operations provided by the application interface.

Administrators are allowed to create and modify the object types that are used to represent the basic organizational components, such as project, department, and so forth, which are needed for the structural modeling of the organizations. Users are allowed to extend the schema by definition or sub-typing of types that are relevant for their local or cooperative work. For example, these are types for the storage of addresses or for the representation of potentially shared working resources, such as tools, notes, or project documents.

The Organizational Information Browser

The organizational information browser provides user access to organizational information. Three major patterns of cooperative work are supported. First, it allows access to and multimedia presentation of organizational information. Second, the interface integrates different communication media to support ad-hoc communication. Third, in combination with a task management system, it provides means for the planning, instantiation, and coordination of cooperative tasks.

Cooperation requires information about the cooperating partners. This ranges from simple address and technical reachability information to their organizational context, which helps to choose the right patterns for communication and cooperation. It is furthermore very comfortable, when the resources of communication are integrated and can be accessed in the same way. This section describes how that information is presented by the interface and how it can be accessed.

The interface allows browsing and searching for organizational information and tracing of organizational relations via a graphical window interface. As well as text information, the interface is able to present different media, which are represented in the information base: graphics used for maps and the presentation of organizational hierarchies, relations, procedures, and rules

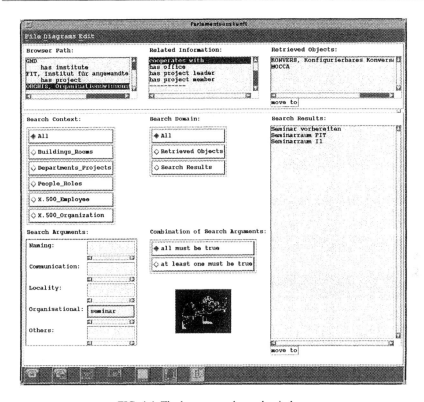

FIG. 4.4 The browser and search window.

and photos (people, groups, buildings, rooms), audio (explanatory text), and video (video demonstrations of software, presentation of public services, etc.).

Normal user interaction starts with a window that provides browsing and querying functionality, as well as means for an easy switching between both search methods (Fig. 4.4).

A set of predefined windows for the display of particular object types and their organizational interrelations has been implemented. In addition, for those that don't have a special presentation, a generic window is displayed that is automatically generated from the object's type information. This reacts flexibly on model extensions done with the object model designer presented earlier.

The whole system is realized as a hypermedia interface. Thus, whenever information is displayed that refers to another information object, it can be immediately retrieved by a user action. That allows manifold ways to access and browse through the organizational information, but it also expresses the various relationships that exist.

Using the mapping information provided as type information for each object, the user interface is also able to display objects that have been read from the X.500 directory. This is useful, for example, when an international project description contains references to members that are not stored locally but are represented as entries in the X.500 directory. Thus, the administration of that information is done remotely by that person, while we still have access to it. This reduces redundancy and guarantees actuality. This external information object is viewed like an internal one. Of course, the user sees a difference in the richness of the data, because X.500 doesn't provide the same amount of data and relations as our system.

For getting an overview on an organizational object and its relationships a graph can be displayed. This is typically an organigram that shows an object in its organizational context. Figure 4.5 shows a graph for an institute. The graph shows the research groups and projects of this institute, as well as the other institutes of the organization. It is generated interpreting the object types graph description. The graph can be used for further browsing, by selecting any of its entries the appropriate object is displayed.

When pictures or maps are used to represent information, they can be used for browsing, too. Linking a picture object by special relationship objects to other objects, areas of a picture become sensitive, so that additional information, such as a more detailed map, or text information is displayed when this area is selected. Audio information can be used to give additional online description.

For each object that is displayed, a simple white board functionality is provided. This allows users to communicate on information they have found in the system. For that purpose, comments can be patched on each object (similar to yellow Post-it notes). These comments can be viewed, added, and modified by all users. It can be used to leave useful experiences or to express problems for other users who look up the same information. It can also be used as by a group of users who start a discussion about an information object, such as about possible extensions on a service that is described, or about informal work-arounds for organizational procedures. With that functionality a communication and discussion tool is directly integrated with the context of discussion, that is, with the information and its organizational context that caused the discussion.

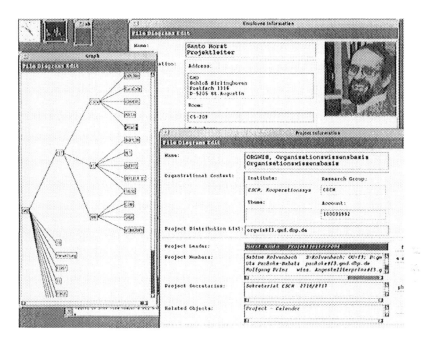

FIG. 4.5 Windows displaying partial information about a project, an employee, and a graph that displays the organigram for the project and its supervising institute.

Cooperation support systems and their user interfaces cannot be treated as closed applications. Integration and interworking with other applications must be possible (Engelbart, 1990). In our system this has been achieved via object adapters for external applications or data. By that technique we have integrated calendars, document editing systems, and so on into our organization browser. This allowed us to turn the information system into a kind of general desktop interface that groups the working resources of a user according to the user's organizational context (private and project calendars, project papers and documents, etc.).

Support for Cooperative Tasks— Application by a Task Management System

TOSCA provides means to describe how tasks or procedures can be carried out in an organization or in a group. This is represented in task template objects according to a task management model that has been developed for the assisting computer (see section 4.1). A task is described in an outline format. This allows the description of major and subordinate tasks. For each single task it can be specified who can support that task or who is responsible to carry it out. Furthermore, resources can be associated to each task, such as documents, forms, calendars, and so forth. This is done by appropriate relationship objects. These are described user specifically. Each user gets individual information about the people who are responsible or the forms that are valid for that user. Thus, TOSCA represents abstract templates that are interpreted and individualized on retrieval.

We stress that the task templates are understood as resources for users to develop their own plan. They are not intended as a prescription for how a cooperative task must be carried out (Robinson & Bannon, 1991).

Task lists can be interrelated, so that users are informed about alternatives or related templates. This increases the visibility of organizational procedures. The white board functionality can be used to comment on work-arounds or experiences one has had in carrying out a task.

Although this information is already very useful as a resource to initiate and carry out a cooperative task, it becomes more useful when it can be transferred into a system that supports its coordination. That integration has been realized within the Assistant for Cooperative Work (ASCW) (Kreifelts & Prinz, 1993), the cooperation support component of the assisting computer. Users can export a task template from TOSCA and then import it into their personal task list. This is convenient for routine task descriptions and it helps when the user carries out a task for the first time. Then the distributed execution of that task is supported by the task manager. In the further process TOSCA is used for address lookup, to resolve role descriptions when administrative offices are involved, to look for substitutes, and so forth.

Architecture

TOSCA is composed of two major components: an organizational information server and the organization information browser (Fig 4.6). The server stores and manages the organizational information objects and relations.

The server is realized on top of a distributed object-oriented database (Ontos). Access to the X.500 world is provided by an integrated directory user agent. All requests for external information are forwarded to the X.500 service. References from the organizational information to X.500 information are automatically resolved. Entries retrieved from X.500 are translated into

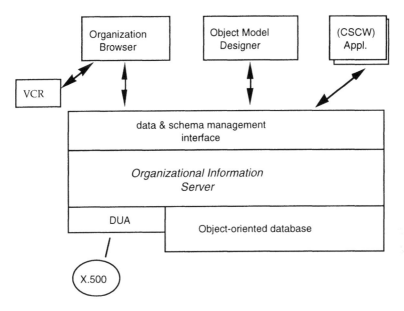

FIG. 4.6 Architecture of TOSCA.

the internal object schema using the mapping information that is supplied for each type.

The server provides a data and schema management application interface to applications. These interfaces are used by the organizational information browser, the object model designer, and by other communication and cooperation support services.

The whole system has been implemented in C++, using GINA (see chapter 5, section 5.1) as an interface toolkit, and Quipu (Kille, 1989) for the X.500 directory components. The server contains currently approximately 750 organizational objects and 600 organizational relationships that are used to represent parts of GMD.

Conclusion

This section presents the motivation, design, and realization of an organizational information service for the support of CSCW applications. We believe that such a service is a fundamental service in the framework of an assisting computer that provides common services to other applications and serves as a helpful assistant for users in their cooperative work. To summarize:

- The system allows the representation of the organizational context for the support of:
  - applications for cooperation support, and
  - users as an information and cooperation support service,
- It provides and integrates different communication media,
- It represents task descriptions as resources for cooperation support of users and applications,
- It provides a visible and user-tailorable object model and thus allows an adaptation to various organizational settings,
- It provides an integrated access to the internationally standardized X.500 directory service,
- In order to provide the best possible representation of information, the system is capable of handling multimedia data.

The application domain for such a service is larger organizations as well as the support for interorganizational cooperation, which becomes more relevant for CSCW in the future. That requires scalable systems on the underlying support and application level. Our design decisions to realize a tailorable, flexible object model, to integrate X.500 access, and to base our implementation on a distributed system make the system scalable to a large extent. In addition, its use for other applications will also simplify their scalability and technical integration into an organizational setting.

The concepts of the organizational knowledge assistant presented in this section will be further exploited within the POLITeam project (Hoschka et al., 1994).

4.4 DYNAMIC INTERFACES FOR COOPERATIVE ACTIVITIES

Markus Sohlenkamp, Thomas Berlage

The success of computer-supported cooperative work depends to a large part on the quality of the user interface. User interfaces for cooperative applications are demanding because the environment is no longer static and the user has to be aware of other people's actions. This is a significant departure from the single-user desktop metaphor. New metaphors as well as innovative techniques are required to effectively present the complex working environment on a small computer screen.

The DIVA (Dynamic Interfaces for cooperatiVe Activities) project has identified the following three key problems to improve the interfaces for cooperative applications:

- A new metaphor is needed for cooperative work. The conventional desktop metaphor is no longer sufficient, because the private desktop must be expanded to a space of cooperating people.

- The concept of time becomes more important. In particular, users must be supported by a history of activities to be informed of other people's actions.

- Cooperative activity must be visualized. New mechanisms, for example animation techniques, are required to provide the level of awareness necessary for successful telecooperation.

Our approach is twofold: on one hand, we extended our application framework GINA (Spenke & Beilken, 1990) to provide generic support for the development of shared applications. On the other hand, the project has developed a version of the "virtual office" metaphor that represents the environment as people moving between rooms in a virtual building. This special application (also named DIVA) tries to mimic the social behavior in a real office on the screen to facilitate communication and cooperation between users (Sohlenkamp & Chwelos, 1994).

The GINA application framework was originally developed to allow the implementation of graphical, interactive user interfaces in an object-oriented fashion. It incorporates a command mechanism to support the easy implementation of an undo/redo feature by using a linear history. By modifying the history mechanism to allow nonlinear history trees, undo and redo of selected, isolated commands, and the transmission of commands to other applications, generic support for the implementation of multiuser applications was integrated into the framework. Applications developed with GINA support synchronous and asynchronous cooperation as well as different coupling modes between the cooperation partners.

The DIVA application uses the features offered by the GINA framework to implement a virtual office—an environment in which to communicate and collaborate with other users. The emphasis is not on a realistic display of a real office, but rather on the modeling of different forms of communication and cooperation in an intuitive way. Objects included in the virtual environment are rooms, people, desks, and documents, displayed in a rather abstract form as individual icons that can be manipulated in much the same way as the icons on the usual graphical desktop for single users. Most other objects existing in real offices are not represented because they are not important in the context of office communication.

The system supports different levels of contact between users—from a brief look into a virtual room to get an idea who is inside, up to entering the room and establishing full communication channels. In the same way, users can signal different levels of communication willingness. For example, they can lock their room to prevent any contacts from other users without explicit permission.

Shared editing sessions are established by putting documents on desks within a virtual room. The coupling mode of a session is determined by the position of the users: if they are sitting at the same desk, they work in tightly coupled mode, while loosely coupling is selected if they are located at different desks.

There are several systems that use similar approaches. The Xerox Rooms system (Henderson & Stuart, 1986) is based on virtual rooms but restricted to single-user, noncooperative environments. Cruiser (Root, 1988), Polyscope, and VROOMS (Borning & Travers, 1991) support communication without offering support for cooperation. VOODOO (Li & Mantei, 1992) supports continuous communication based on the open-office metaphor but also includes only very limited support for cooperative applications. The cooperative Sepia system (Haake & Wilson, 1992) supports both communication and cooperation, but is limited to a single application. The DIVE system (Fahlen et al., 1993) also models aspects of the real office in a virtual world, but is based on a virtual reality approach.

Virtual Environments

Although the development of multiuser applications is greatly facilitated by the GINA application framework, there are some issues that are not addressed by such a development tool. One of these problems is the integration of multiuser applications into a working environment. This is the reason DIVA was developed: It provides an environment that simulates several aspects of a real office, especially those concerning cooperative work, and integrates them with the concept of shared applications.

Many of the potential benefits of a virtual environment supporting communication and cooperation are already described in Root (1988). One of the advantages is that work groups can be separated both in space and in time. Separation in space becomes possible because there is no need for a work group to share a common office building. All communication is done via a computer network, allowing arbitrary distances between communication partners. For the same reason, reorganisation of work groups is facilitated because there is no need to move anything physically. Only the assignment to a virtual environment has to be changed. Separation in time becomes possible by supporting asynchronous cooperation, navigation in the command history, and the merging of different versions of a document.

Furthermore, a virtual environment offers a great degree of flexibility. Rooms do not need to have real-world counterparts. As an example, it is possible to create task-related rooms instead of the usual person-related ones—for example, project or documentation rooms. Users can arrange their virtual office according to their needs with regard to the general layout and the selection of rooms and other users they want to see. For example, an individual placement of rooms is possible to allow rooms of important communication partners to be placed close to one's own room.

It is important to support different forms of communication and cooperation present in real offices. A virtual office application should offer the possibility to:

- Establish communication channels between users, namely, video and audio connections.
- Display information about other users, such as their availability, communication willingness, or activities, but respecting privacy issues.
- Start shared editing sessions via multiuser applications.
- Join and leave multiuser sessions.

- Choose different coupling modes for shared sessions.
- Create, edit, copy, and delete documents.

All these services should be offered in a transparent way, such that the user should not have to know anything about the underlying mechanisms and applications that need to be started. The access to a service should be intuitively deducible from its real-world counterpart.

We believe that it is important to abstract from the characteristics of a real office: Only the essential mechanisms and an intuitive access to them are relevant. Therefore, it is not useful to model all the features of a real office, especially because they include many shortcomings and drawbacks that would make both the design as well as the use of the virtual office unnecessarily complex. For example, a realistically modeled telephone would still need a number to be dialled to make a call. There are clearly much better ways to realize this on a computer network. The same is true for the modeling of the actual office building: realistic displays, such as of hallways, may only complicate the navigation in the virtual world. Therefore, DIVA does not use the virtual reality approach as do other systems such as DIVE (Fahlen et al., 1993) but relies on an extension of the standard desktop metaphor instead. The general difficulty is to find a compromise between a recognizable model and the necessary abstractions.

The DIVA Virtual Office

To keep the user interface simple, DIVA basically offers only four classes of objects in the virtual world. These are people, rooms, desks, and documents. Additional minor elements are briefcases, stick-on notes, and a trash can.

People represent the users of the DIVA system and are implemented as small snapshots with a name beneath. They can be moved by their owner to change their virtual location and thereby determine what they are doing and whom they are in audio/visual contact with.

Documents represent the artifacts people work on in the virtual office. They are similar to documents in the standard single-user desktop, but have additional properties, such as change status and current activity status, which are needed to adequately support group work. They differ from real documents in that they may be in two or more places at once, for the sake of convenience. The multiple copies in different places are simply access points for a single shared artifact.

Desks serve a variety of purposes: they are the arena for work within a room, they control the coupling mode of a cooperation, and they can be used to preserve working context. A special form of a desk is the briefcase, which contains private documents.

Rooms are containers for people, desks, and documents. They also control the audio/video communication status of users. Just as people located in the same real room are able to see and hear one another, so too can people in the same DIVA virtual room hear and see each other; when a DIVA user enters a virtual room already occupied by one or more users, audio and video links are established between the newcomer and the other occupants. Rooms also serve to indicate availability and communication willingness: they can be in different states, providing different levels of access and visibility of their inhabitants. Rooms can be used as private offices, public meeting places, or special-purpose spaces.

Rooms themselves are contained in the DIVA virtual office environment. Users may customize their virtual office by selecting their set of potential cooperation partners and placing the rooms as they like. A glance at the rooms contained in the virtual office shows who is inside each open room.

People move through the virtual office, from room to room, by dragging their icon. They may glance into rooms that are not open, and they enter a room in order to do work or establish contact with others inside. Once in a room, the desks and documents in it become visible. By moving shared documents to a desk in the room, users may work together. People working at the same desk are engaged in a focused collaboration, using a tightly coupled editing mode, while people working on the same document but at different desks are less focused and use a loosely coupled mode.

Working with DIVA

A typical DIVA session is illustrated in Fig. 4.7. The virtual office, shown from the point of view of user Markus, is displayed in two main windows. The first window (in the background) contains the virtual office itself, and the second (in the mid ground) shows the virtual room that the user is currently in. The virtual office window displays the DIVA rooms and the people present in the virtual office. It indicates the locations of the users and displays their movements from room to room as animations. Users organize the DIVA rooms as they like by simply dragging the rooms with the mouse.

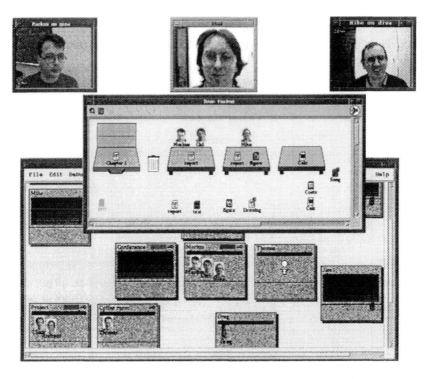

FIG. 4.7 The DIVA system.

A glance at the virtual office window provides a broad level of awareness of coworker activities: Markus, Cici, and Mike are together in Markus' office; Claus and Andreas have met in the project room; Greg and Thomas are each alone, but available for contact; the people in the "Conference" room would like some privacy; and user Jim does not want to be disturbed, as indicated by the lock on his DIVA office.

In contrast to the standard desktop, the virtual office display is not static, but changes dynamically with the actions of other users. For example, the room icons are used to reflect the room status and can change any time according to the needs of their owners. They can be open, closed, or locked—displaying a different graphical image for each state. Persons inside an open room can be seen by others at any time. The blinds of a closed room open if someone tries to enter it or comes near, but the visitor will likewise be seen by the users inside by displaying his icon in front of the room. This symmetrical relation is equivalent to looking through a door in a real office—this, too, cannot be done without the persons inside noting it (Root, 1988).

Finally, a locked room effectively disables users from entering it—unless they are in the list of key owners of the room. For those users, a locked room works the same way as a closed room. Every key owner can change the status of a room via a corresponding menu entry—for example, a room can be locked after all participants of a meeting have arrived, to prevent any other persons from joining.

The icons of the users inside a room are displayed within the icon of the room, giving the visual impression that the users are in the room. Therefore, if most of the rooms are kept open, a quick look at the display provides a good overview of the activities in the office: which users are in which rooms, who is talking with whom, where meetings take place, and so on.

If a user enters a room, an acoustic signal is given to all other users in the room to inform them of the presence of a new user. This is followed by the establishing of video and audio connections.

Because the movement of other users is displayed as an animation, there is also a visual indication of the overall activity in the office. On the other hand, every user can switch off these displays if they are not needed or considered to be disturbing.

The second DIVA window, labeled "Room Markus" in the example, is the virtual room window. It reveals the contents of the room that the user is currently in. In addition to the people who are in the room, the desks and documents in it are shown. Awareness information is also conveyed by this window: Markus, Cici, and Mike are all working on the shared document "report"; Mike is also editing a drawing named "figure"; the documents "Song" and "text" have changed since Markus looked at them; and there are two notes for Markus in the room. Working at the same desk, such as Markus and Cici are doing, indicates focused work and results in a tightly coupled synchronous editing session. Although Mike is also editing the same document, he is at another desk so his work is independent and he is loosely coupled with the others. In this manner work on the shared artifacts in the room focuses around the desks in the room, as it usually does in real offices. On the third desk, a spreadsheet remains open for future work, illustrating the use of desks to preserve working context.

Finally, the small windows shown in the example are live video images of the people in the room.

Awareness

To effectively support cooperation between different users, it is crucial to provide awareness—knowledge about the activities of other users (Dourish & Bellotti, 1992). The DIVA system supports awareness on different levels. As described in the previous section, many facts about other users can be observed directly in the virtual office window, including their location, communication status, and availability. The virtual room window offers additional information: who is working with whom on what documents. The status of documents is reflected by the background color of their icon: documents that are in use by others are displayed in red; those that have been changed by others are in green. To inform the user of rooms containing such documents, a bar in the corresponding color is displayed on top of the room icon in these cases.

During cooperative activities, multiuser applications developed with the GINA framework offer additional information about the activities of other users. For synchronous cooperation, changes in a document—that is, the results of some user commands—are immediately reflected on all displays. Furthermore, in the tightly coupled mode, the way a command is invoked is also made visible to all cooperation partners. This includes the animation of menu selections, the popping up of dialog boxes, the scrolling of windows, and graphical feedback for direct manipulative commands. Additionally, every user is associated with a color, and every application displays a list of connected users and their respective colors. Because all operations of a user are animated using his or her associated color, it is easily recognizable who invoked a command.

For asynchronous cooperation, every user can get an animated display of the document history—either continuously or step by step—to be informed about the changes of other users. Again, commands related to a certain user are drawn in his or her associated color, both in the animation as well as in the graphical representation of the history tree, allowing an easy match between document changes and users.

The combination of DIVA with shared applications offers even more possibilities: In addition to providing an audio and video link to support informal communication with the cooperation partners, it also displays temporarily flashing lines connecting the video image of a user to the cursor position while he or she is working. This unobtrusive form of information can be noticed subconsciously by other users, assisted by the fact that the video images of the cooperating persons are located in different corners of the screen.

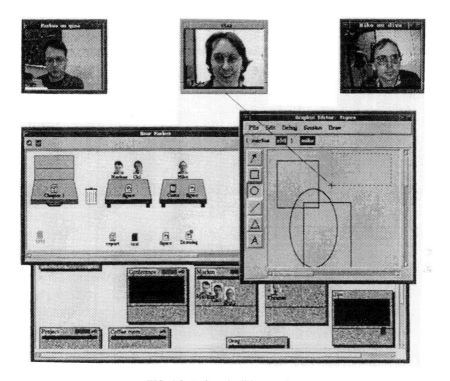

FIG. 4.8 A shared editing session.

Figure 4.8 shows a shared editing session with DIVA, again from the point of view of user Markus. Three users are jointly working on the drawing "figure". Below the menu bar of the graphics editor, the names of the participating users, their associated colors, and their coupling mode is displayed. Tightly coupled groups are displayed in brackets. All users can make changes to the document which are immediately reflected on the screens of their cooperation partners. While user Cici is drawing a new rectangle, a flashing line linking his video image and mouse position makes his action visible for user Markus.

Technical Background

DIVA is based on the application framework GINA. The original GINA supported the development of single-user applications and incorporated a history mechanism that allowed an unlimited undo and redo of commands. By extending this mechanism to support a nonlinear history and a selective undo/redo feature, it turned out that GINA could be used as a generic application framework for multiuser applications as well (Spenke, 1993). It provides support for different forms of coupling between applications: from strict WYSIWIS (What You See Is What I See) over common document state to complete decoupling. Furthermore, the system allows smooth transitions from one coupling mode to another.

The mechanism is based on command objects that are used to encapsulate all user actions (Berlage & Genau, 1993). In the coupled cases, command objects are broadcast by the originating application to all connected applications and are replicated locally. Virtual time stamps are used to detect conflicts, which are resolved by reordering the conflicting commands. What types of commands are transmitted depends on the coupling mode. If only the document state is shared, it suffices to transmit state-changing commands. Otherwise, additionally commands that change the view on the document, such as the scrolling of windows or the selection of objects, have to be transmitted. Of course, there is no need to transmit any commands in the decoupled case. The problem here is the merging of different document states if a closer coupling mode is selected later. This is realized by regarding the different states as different branches of the history tree and by appending one branch at the end of the other by means of the selective redo feature. Graphical displays of the history tree help the user to navigate through different versions of a document.

Video and Audio

DIVA uses a video and audio link to provide communication between users. For the transmission of audio signals, most workstations offer hardware support (microphones and speakers) to record and replay sound samples of at least telephone line quality. This is enough to support verbal communication. For the video link, special hardware is needed. In our case, this consists of a video camera mounted on top of the computer monitor and a special video board that allows the display of a digitized video signal in a window using hardware JPEG compression of images. This configuration allows the usual face-to-face conversations in a video conference.

The audio connection is essential to permit at least a minimum of communication. In contrast, the video information is not absolutely necessary in a shared session. Therefore, machines without the special video hardware can also be used to participate in the virtual office by transmitting no picture information at all.

Video and audio connections are automatically established between all users sharing the same room. Video images can be manipulated and adjusted individually as well as switched off totally. The image others receive from an individual user is controlled totally by this user: Besides switching off the camera, the user can also disable the transmission of video data temporarily—in this case, a predefined still image is displayed to the other users. Because the own image is reflected for each user, it is possible to constantly supervise what picture is sent to others. In an analogous way, users control their audio connections; they also can be turned off individually, for example, to answer a call on the phone.

Technically, the transmission of video and audio data is accomplished by the use of two separate processes: a video and an audio server. Thus, DIVA relies on three server processes to do its work: the X server to handle user input and display graphics, the audio server to perform audio input and output, and the video server to display the digitized video images. Both video and audio data are transmitted via standard network connections (e.g., TCP/IP over Ethernet). Figure 4.9 shows the different parts comprising the DIVA system for two connected machines. X, AF, and VS denote the X server, the audio server, and the video server, respectively. Arrows show connections between components, and dashed arrows symbolize potential connections to additional machines.

The AudioFile system (Levergood et al., 1993) is used as the audio server: It is device independent, network transparent, and allows multiple applications to share the same audio hardware. The basic mechanisms for audio input and output resemble those of the X window system for graphical input and output.

Although the device independence is only a potential benefit for DIVA, network transparency and the fact that multiple clients can share the same hardware are crucial: they allow the mixing of different sound input channels, for example, the voices of several communication partners all speaking together. Furthermore, sound effects can easily be mixed into an audio conference to provide additional cues to ongoing activities. Audio conferences are realized by special AudioFile clients that collect audio input on every machine that runs DIVA and replay it on connected machines.

For the video conference, a video server, based on X and the special extensions for the video hardware, is used. It is an X client with two basic capabilities: It displays a digitized video image in a given window, and copies the contents of arbitrary windows to other windows located anywhere on the network. For the copying process, the JPEG compression built into the video hardware is used to maximize the transmitted frame rate.

Once started, the video server performs its duties automatically by permanently displaying the video image from a connected camera in a given window. Requests from the client program to start or terminate the copying of the contents of that window—and thus to start or terminate a video conference with another user—are given by special X events that encode the necessary parameters. The frame rate of the video transmission is adjusted automatically to the network load to ensure high audio quality and an acceptable response time of the shared applications.

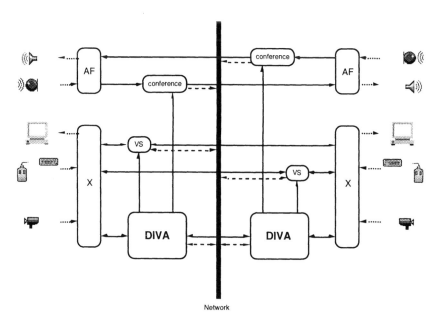

FIG. 4.9 DIVA: schematic overview.

Future Work

Many questions regarding the design of the user interface of the virtual office still remain open to further research.

A better approach for the graphical representation could be the usage of three-dimensional techniques. Because of the reasons discussed earlier, this does not mean virtual reality, but the use of simplified 3D images of objects displayed from the point of view of the user, rotating, moving, and zooming according to the user's movement in the virtual world. This would allow a more compact display of information, and 3D effects could enhance the illusion of moving in a virtual world and thus facilitate the intuitive understanding of the mechanisms of the virtual office. The usual input devices should be used to navigate in 3D space to prevent the necessity of additional hardware.

The use of animation to convey information about the virtual world should be extended. Apart from showing the movements of other users, animation could be used to illustrate changes of other objects as well. This applies especially to documents: creating, opening, moving, or deleting a document should be animated to give other users a cue to what is happening. The animated display of the history of a document provided by the GINA framework should be integrated in a better way into the virtual office. People joining a conference could automatically be given an animated overview of changes made to the document while they were absent.

In the same way, the use of sound effects could be extended. An example would be creating the sound of murmuring when passing the door of a room where a conference is held.

A potential problem that could prevent widespread usage of the system is the inability to integrate existing standard office applications into the environment without losing much of its functionality. Therefore, we plan to transfer the ideas described in this section into an industrial setting with standard applications and see how the concepts can be applied here.

Conclusion

The system described in this section offers a lot of promising features and potential benefits. These are mainly concerned with the improvement of communication and cooperation between work groups, individually customizable working environments, and integration of shared applications.

Although further studies are necessary to determine whether users really accept such a system and use it for their work, first tests indicate that the system is in fact very easy to learn and to use.

5 Methodological and Tool Projects

5.1 FRAMEWORK FOR GRAPHICAL USER INTERFACES

Thomas Berlage, Michael Spenke

The AC was planned to consist of many components—the assistants—with consistent user interfaces designed according to uniform guidelines. Therefore, a common development base for all projects was defined: the UNIX operating system, the X Window System, the OSF/Motif graphical user interface, and LISP/CLOS (Common LISP Object System; Keene, 1989) and C++ as programming languages. However, it soon became clear that this decision alone was not sufficient to guarantee consistency. The application framework GINA (Generic INteractive Application) has been created in order to define guidelines and to incorporate them into a software tool to be used by the projects implementing assistants.

GINA employs object-oriented techniques to implement the common functionality of the assistants as a reusable class library. The common code is mainly concerned with user interface issues, but other aspects like loading and saving documents are also handled. New applications are created by defining subclasses of GINA classes and adding or overriding methods. Only the application-specific differences to the standard application have to be coded.

The power of a generic application can be explained by a metaphor: Using an interface toolkit is like building a house from scratch, whereby a lot of guidelines have to be followed. Using a generic application is like starting with a complete standard house, already following the guidelines, and adding some specific modifications.

The concept of an application framework has some advantages that are of special importance in the context of the AC project:

- User interface guidelines are not only written on paper, but can be implemented in software. Thus, a uniform interface can be guaranteed to a large extent.

- The implementation effort for an individual assistant can be considerably reduced.

- Because there is a common layer of software for all assistants, better integration and cooperation among the AC components are possible.

- Future extensions and innovative interface features can later be incorporated into existing assistants. For example, the latest GINA version supports multiuser interfaces.

GINA is based on concepts known from MacApp (Schmucker, 1986) and ET++ (Weinand et al., 1989). There are a number of other frameworks for user interface construction, such as Interviews/Unidraw (Vlissides & Linton, 1990), Garnet (Myers et al., 1990), and Picasso (Rowe et al., 1991), but most of them focus on the lower layers and start with their own toolkit. GINA takes a different approach by encapsulating an existing toolkit and concentrating on the higher levels to lay the foundation for more "intelligent" support of the human–computer communication.

When the project started in 1989 the idea of an object-oriented application framework was not completely new. For example, the first versions of Apple's MacApp were available. However, no off-the-shelf software tools were available for the system base of the AC, which is very much oriented toward standards. Moreover, the concept of an application framework was extended in many directions (Berlage, 1993a):

- GINA encapsulates the industry-standard OSF/Motif toolkit and integrates it into a coherent object model with the rest of the application framework.

- GINA provides *direct manipulation objects,* which couple application-specific graphics with configurable behavior and undo support.

- GINA maintains an unlimited history of the user interaction. Animated replay of the history is available as a help facility and a learning aid.

- Selected parts of the history can be independently undone.

- GINA implements the *tap model,* which allows changing the interface without changing the underlying application code.

- In its Common LISP version, GINA provides an incremental programming environment. There is special support for error handling in an interactive application. The GINA reference documentation is automatically generated from the source code and is also available for online access.

- GINA includes an interface builder to interactively define the interface. The interface builder not only includes features that make using Motif easier, it also integrates its output well with the rest of the framework.

GINA has not only been used inside GMD to implement several assistants but is also publicly available (Spenke et al., 1992a, 1992b). It is used by several hundred users in research institutes and commercial companies all over the world. CLM (Common LISP Motif)—a component of GINA—is now generally accepted as the standard interface between Common LISP and the OSF/Motif toolkit. It is used in several commercial products.

GINA Classes

GINA implements a certain common model on which all applications are based. It is closely related to the Macintosh application model. As the Macintosh shows, the model is sufficiently general to cover nearly all types of applications. GINA does not dictate this model; it simply offers more support for applications that follow the model.

There is exactly one object of class application (or more precisely: of a subclass of class application) for each running GINA application program (Fig. 5.1). The application object holds a list of document objects, one for each open document the user is working with. Each user action already executed for a document is stored in a list of command objects that form the base for undo and redo operations. Document objects themselves are invisible in the user interface; they store the internal representation of the document contents and create a shell to display their contents. A shell is the root of a tree of Motif widgets. The leaves of this tree are widgets (like push buttons or scroll bars) or views (empty widgets where the application can draw using the X Window primitives). Graphical objects within a view are also modeled as CLOS objects.

Here is a short overview of the GINA classes:

APPLICATION: This class constitutes the main program of a GINA application. It contains the main event loop translating incoming events into messages to other objects. At runtime there is exactly one instance of a

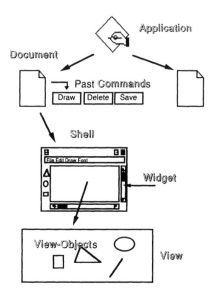

FIG. 5.1   The most important classes of GINA.

subclass of class application, differing from the superclass at least in its name, signature and the type of document objects used.

DOCUMENT: This class corresponds to the documents seen by the user, which are created by the user by a new or open command, and can be saved to a file. The programmer defines a subclass of document adding slots describing the contents of the application-specific document. Furthermore, the programmer has to override the methods write-to-stream and read-from-stream, so that the document contents can be saved to and restored from a file. GINA calls these methods when the corresponding menu entries have been activated by the user. The programmer also has to override the method create-windows. This method creates the window that represents the document on the screen.

SHELL: Shells are top-level windows that can be moved, resized, and iconified by the user (using the window manager). The shell classes used most often are the document-shell, which already contains a menu bar and is used to represent a document, and the classes modal-dialog-shell and modeless-dialog-shell, which are used for dialog boxes.

WIDGET: This is the superclass of all Motif widgets. Special widgets like scroll-bar, push-button, and so on, are subclasses of this class.

VIEW: Views are drawing areas where the contents of a document are represented graphically. The programmer can override the method draw of class view and call the graphical primitives of the X Window System inside this method. GINA calls this method whenever any part of the view has to be redrawn. Views can also react to keyboard and mouse input. To achieve this, the methods button-press and key-press, which are called when the corresponding events are received, can be overridden.

VIEW-OBJECT: Instead of overriding the draw method of class view, it is also possible to install view-objects at a certain position within a view. View-objects have their own draw method defining their individual shapes. The draw method is called whenever a view-object must be redrawn. Mouse events are propagated to the object under the pointer. There are methods to move and resize view-objects. A typical example for view-objects are the objects of a graphical editor.

DIRECT-MANIPULATION-OBJECT: This is a subclass of view-object. These objects can be selected, moved, and resized using the mouse. Methods can be overriden to define individual feedback during dragging operations, nonstandard highlighting of selected objects, and so on. The subclass movable-icon covers an important special case.

COMMAND: The command objects form the base for undoable user actions. Each relevant user input leads to the creation of an object with superclass command, which stores all parameters. The effect of a command is determined by the method doit, which is always overridden by the programmer. Later, GINA may call the method undoit (also overridden), which implements the inverse operation. Therefore, the application programmer has to define a subclass of class command for each kind of undoable command supported by the application.

MOUSE-DOWN-COMMAND: This is a subclass of class command that is initiated when the mouse goes down in a view. As long as the mouse button is held down and the mouse is moved around, an application specific feedback is shown in the view. A typical task for a mouse-down-command is dragging a rubber-band rectangle in a graphic editor. For dragging objects across window and application boundaries, the class drag-command is available.

Dialogue Architecture

Following the Arch model (UIMS Tools Developer Workshop, 1992), an application can be divided into five parts that successively implement a mapping between the application functionality and the user interface (Fig. 5.2).

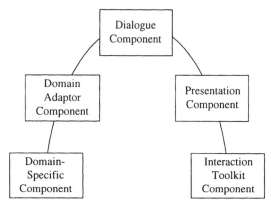

FIG. 5.2 The Arch model of a graphical interactive application.

Figure 5.3 shows which GINA classes are used to implement the components of this model. Starting from the right, the toolkit component provides objects that can be composed into the visible part of the user interface. On the one hand, GINA supports the objects of the Motif toolkit, encapsulated into classes of the programming language used. For application-specific graphics

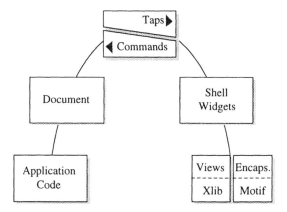

FIG. 5.3 GINA classes in the Arch model.

(such as the drawing area of a graphical editor), GINA provides views and view objects that have to be subclassed by the application programmer.

The interface objects are tied together by the presentation component. The presentation component decides how to present a feature of the application. In GINA, this task is performed by a shell class to be subclassed by a programmer. The code in the shell class creates the composition of objects and connects them input-wise with commands and output-wise with taps.

The dialogue component determines the dynamic behavior of the user interface. In GINA, every input action reported by a toolkit object results in creating an instance of a command class that is collected into the history by GINA. For every different user action, the programmer has to write a command subclass, giving it at least a name and implementing the effect. Most of the commands affect application data; only few of them stay within the interface side.

Commands are not responsible for updating the display. This task is performed by taps and makes the commands relatively independent of changes in the presentation. A tap is a procedure triggered by command execution that maps some part of the application data in the document to attributes of the interface objects. Taps can also dynamically create and destroy interface objects.

The abstract representation of the application data is modeled by a document subclass in GINA. The purpose of the document is to model the application in terms of the document model as found on the Macintosh. The substructure of the document is application specific. The programmer has to define methods to write the document contents to disk and read it back.

The application code is not further constrained by GINA. It may be written in any language, provided the contents and functions can be accessed through the document object.

Interfacing LISP and OSF/Motif

In GINA, we decided to use a commercially available toolkit instead of writing our own. We chose OSF/Motif (Berlage, 1991a) because it provides the widest availability on UNIX platforms. OSF/Motif is based on the X Window System (Scheifler & Gettys, 1986) and the X Toolkit. It consists of a special window manager and a set of interface objects (widgets) such as push-buttons and scroll-bars. The widgets are implemented in the C programming language.

Using Motif poses two challenges for GINA. First, the Motif widgets must be made available in Common LISP, and second, they should be encapsulated as classes in a real object-oriented language (CLOS or C++).

To connect OSF/Motif and LISP, we have implemented CLM, a package that runs the Motif software in an extra Motif server process implemented in C (Bäcker et al., 1992). The LISP application communicates with the Motif server using a special protocol, similar to the X protocol, but at a higher semantic level. Each LISP application running in a single LISP process has its own Motif server process. Low-level interactions such as browsing through a menu or dragging a scroll-bar can be handled by the Motif server, without the need to run any LISP code, which results in good performance at the user interface level.

The LISP application can also directly contact the X server in order to perform graphic output into drawing areas.

Although the X Toolkit Intrinsics, on which Motif is based, implement an object-oriented model with a class hierarchy, the widgets should be encapsulated as objects in the underlying programming language (CLOS or C++) to create a homogeneous environment (Bäcker, 1992). For this purpose, every widget is modeled as an object that solely consists of a pointer to the widget. All methods of this object are passed to the widget. We have implemented the following additional properties in our encapsulation:

• Callback objects to enable callbacks to call methods.

• Enable subclassing of composite widgets to replace standard dialogs.

• Unified model for selection classes, modeling a single value.

In Motif, input is handled by calling a callback procedure from a widget when a widget-specific external event occurs (such as pressing a button). In an object-oriented environment, callback procedures are methods to be called with a specific object. Therefore a callback cannot be a simple procedure address, but is replaced by a callback object that contains the name of the method and the object for which the method should be called.

In an object-oriented system, it is natural to specialize objects by subclassing. This is not possible with encapsulated widgets, because there is no access from the target language to the widget internals. However, especially with widgets, new classes are often a composition of other objects. The GINA widget encapsulation supports creating composite widget classes.

In addition to the main widget pointer, an encapsulated widget has two other widget pointers. For a single widget, they are all identical. For composite widgets, one pointer designates the top widget of the subhierarchy and the other pointer designates the widget where further children may be added. Methods of the composite widget are passed to the appropriate one of the three pointers. This mechanism makes it very easy to define specialized composite widgets.

The widget composition mechanism is also used to define a unified selection model for Motif. Motif provides a number of different widgets to present a selection among a list of alternatives to the user (a selection list, an option menu, a radio button group, radio buttons in a menu). In the first approximation, all these widgets are equivalent as seen from the programmer. Choosing one of these widgets for an interface should be based on ergonomic criteria and should not require changing the application. However, in Motif all these widgets have a different programmer interface.

For example, a radio button group is composed of individual toggle buttons, one for each choice, whereas a selection list is a single widget that accepts a list of strings for selection. In GINA, a radio button group is defined as a composite widget that also accepts a list of strings and creates the appropriate number of buttons itself (even dynamically). The classes for selection come in two versions: one for single selection (1-of-n) and one for multiple selection (n-of-m). To make identifying the user choices easier for the application, every string presented can be accompanied by an arbitrary object or value of the application.

Individual Graphics

Motif and similar toolkits do not provide all necessary interface objects. There is a natural division between standard objects (such as push-buttons, scroll-bars, and menus) and application-specific objects (such as circles, irregular icons, or spreadsheet cells). The former are provided by Motif, and the latter are individually created as GINA subclasses.

The programmer can choose between a procedural and an object-oriented interface for graphic output. Using the procedural interface, the programmer overrides the draw method of his subclass of view. In this method he can call the X primitives which are available as methods of class view. GINA calls the draw method whenever necessary.

The object-oriented interface to graphical output is implemented on top of the procedural one. A view can store a list of so called view-objects. The class view-object is the superclass of graphical objects like circle, rectangle, and line. View objects can be installed at a certain position in a view and later can be moved or resized. They remember their own size and position. The view ensures that an installed view-object will be redisplayed whenever the corresponding part of the view is redrawn by calling the view-object's draw-method.

The graphical object system in GINA is similar to (but sometimes not as refined as) the systems that can be found in InterViews (Vlissides & Linton, 1990) or Garnet (Myers et al., 1990), for example.

Input

In an object-oriented system it is possible not only to define static entities as objects, but also a process. In GINA, every user interaction (from key or mouse press until release) is implemented as a command object. Command objects are used to implement an interaction history with undo support (see next section), but also to describe mouse operations.

GINA defines command classes for mouse operations. For a specific operation, the programmer has to create a subclass. This subclass will inherit a lot of complicated behavior, such as handling autoscrolling or the effect of exposures on the current feedback. The programmer only has to code the behavior that is specific to the operation.

Commands for mouse operations are similar to the interactors in Myers (1990). Command objects are different because a new instance is created for every operation, whereas there is only one interactor for a whole class of operations (the history aspect is missing).

GINA also defines command classes for drag-and-drop operations between different applications. The additional problems here are the identification of potential targets in other applications and the implementation of feedback among two applications. The latter problem is solved by having two parallel command objects in each application that handle the respective parts of the protocol for their own application.

History

GINA provides object-oriented support for storing the dialogue history and using it in different ways (Berlage & Spenke, 1992). The history is used to implement an advanced form of undo, which is very important in direct-manipulation interfaces (Berlage, 1993b). Furthermore, the history is used as a help facility.

Unlike MacApp, GINA stores an unlimited history of commands. Whenever a command is executed, it is added to the history. "Undo" and "Redo" are available as menu entries. They are meta-commands and thus do not appear in the history. The history consists of two parts: the undo list of executed commands and the redo list of undone commands that can be redone (Fig. 5.4). Undo calls "undoit" on the last command in the undo list and prepends it to the redo list. Redo calls "redoit" on the first command of the redo list and appends it to the undo list.

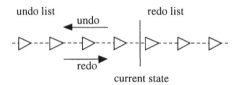

FIG. 5.4 Commands (shown as triangles) in a linear history.

This model is called "linear undo" (Archer et al., 1984). As an alternative access to the history, GINA implements a "history scroller" (Fig. 5.5) that can be used to scroll backward and forward in the history, undoing or redoing the commands on the way.

FIG. 5.5 The GINA history scroller.

In the linear undo strategy, a user cannot undo isolated commands from the history. In GINA, we have developed an extension of the linear history model to allow selective undo and redo of isolated commands. Instead of editing the history, the desired command is copied ("cloned") and appended to the history. The copy is then executed as a new command. This model is clean and easy to understand and ensures that existing parts of the history are never modified (Fig. 5.6).

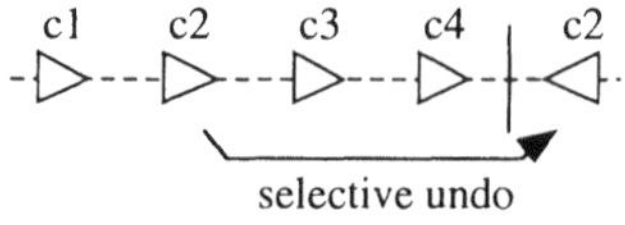

selective undo

FIG. 5.6 Undoing an isolated command from the history
by cloning it and marking the copy as "reverse."

It is not always possible to undo or redo an arbitrary command from the history tree—for example, if the affected object no longer exists. To support this mechanism, every command class implements two application-specific methods "selective-undo-possible" and "selective-redo-possible." If they return "False," the command is not available for selective undo/redo. Furthermore, a method to clone a command object must be implemented for each class.

There are two new commands for the user: "Selective Undo" and "Selective Redo." Both present a list of commands that can be undone or redone by copying, that is, those that have returned "True" from their applicability check (Fig. 5.7). The commands to undo are selected from the undo list, because these are the commands currently in effect. The commands to redo are selected from the redo list and from all the cut-off history branches, because these commands have already been undone.

In these lists, commands are described by a longer string that tries to include all the parameters of the command and the objects affected. This description is also generated by a method of the command class. Because the list will usually be quite long, the user must be allowed to select a subset of commands. Most probably the user is interested in a command affecting a certain object or with a certain name. For example, all text changes of a specific paragraph can be selected.

A simple string search mechanism is sufficient for selection. The user can specify a number of substrings implicitly connected with AND, for example, "Change Text" and "%<Paragraph 05541>". Names of commands or objects can conveniently be generated by clicking on the object or a command-executing button.

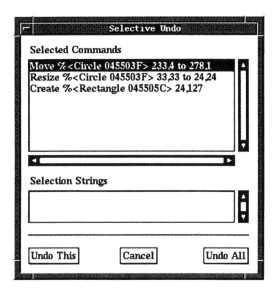

FIG. 5.7 Selective undo dialog.

History Animation

In direct-manipulation user interfaces, users often have the problem that they do not know exactly what they have just done (e.g., when keys have accidentally been pressed or a mouse target has not been hit exactly). These questions can be answered by the command history (Baecker & Small, 1990).

However, when browsing through the history using undo and redo, only the state-changing effect of the command is shown, not the detailed interaction that was necessary to submit the command, such as the feedback when moving objects. To obtain the additional information, the *interactor objects* (Myers, 1990) that control the interaction (such as the feedback during mouse motion) and record any necessary details for replay are also stored in the history. The interactors are application independent and can be used to replay the exact interaction. Like a film, animation makes sense only in the forward (redo) direction.

Mouse and keyboard are normally operated by the user and must be simulated during the animation. A second "simulation mouse" shows the pointer position. Mouse clicks are simulated by darkening buttons shown on the mouse image, supported by sound effects. Key presses are simulated by popping up an image of the key for a short time. We are still experimenting with these effects, which is possible because they do not affect the programming interface of the existing applications.

For a complete history animation, all user interactions must be recorded. This creates a problem with operations such as scrolling that normally would not appear as undoable commands because they do not affect the information content. In GINA, those interactions are implemented as commands with a flag "causes-change" set to "False."

In GINA, the history is permanently stored together with the data manipulated (the document). In this way, the history can also be replayed by others, for a variety of purposes:

- Users can send their documents as bug reports and the programmer is able to reconstruct the history that led to the bug. Support staff can analyze documents of novice users in case of trouble and send them back with a demonstration of a solution.

- Users can also keep the history of a quick demonstration given by an expert and safely reconstruct it later.

- Program features can be demonstrated by creating a document that uses these features and users can watch for themselves how the result has been achieved.

Because the history usually refers to other application objects, it is only valid in conjunction with a specific document state. Therefore, the history cannot be separated from the document.

For help purposes, it is possible to annotate the history with explanations. We have implemented a simple, recorder-like extension of the history scroller that is able to record voice annotations and descriptive texts for every history step. The user can scroll to any point in the history and start the animation from this point. The recorded sound is played back and the text annotations are shown during the animation.

Output

It is very important to decouple the output behavior of a graphical application from its data manipulation. If display updates are mixed with data manipulations, it is very difficult to change the interface, because code must be changed in a number of different locations. This is especially severe if there are multiple views on the same data or if interfaces change dynamically.

A number of schemes have been devised for this purpose, most notably the Model-View-Controller (MVC) model of Smalltalk-80 (Krasner & Pope, 1988). Constraint mechanisms (Borning, 1986; Myers et al., 1990) can be used, too. These models are based on a *change-propagation* protocol where a change in an application object notifies a number of anonymous dependents. This protocol requires some support from the application objects. GINA uses a different technique, based on *taps* (Berlage, 1992b), that does not require support from application objects.

A tap is a partial mapping of application objects to an interface object. Taps are triggered by user actions, not by the application objects that change (Fig. 5.8). A tap specifies a procedure written in a general-purpose programming language and a set of commands after which the procedure must be evaluated. A command in this context can be any user interaction reported by a callback, such as clicking a button or dragging an object.

Taps are associated with interface objects, but there is considerable flexibility for defining the mapping. An object usually has several taps that control different attributes. The same attribute may be calculated by multiple taps for different commands.

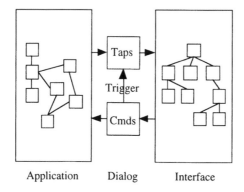

Application   Dialog   Interface

FIG. 5.8 Separating input and output.

The advantages of taps are:

- Interfaces can be added or modified without changing the application code. Additional interfaces and their taps can even be added at runtime (e.g., for cooperative applications).

- The application code contains no pointers to the interface, so existing application objects can be used unchanged. This is especially important for databases.

- Taps can easily be implemented in conventional object-oriented languages, such as C++ (compared to constraints).

- Taps can be specialized according to the different user commands. There is no need to classify the changes in the application (such as in MVC).

The main disadvantage is that taps are triggered more often than necessary, because all taps that are potentially affected by a command must be evaluated. A number of measures are described next that make the performance degradation tolerable.

The Programming Environment

The GINA version implemented in Common LISP greatly benefits from the interactive nature of typical LISP implementations. In particular, methods and functions can be changed, recompiled, and substituted into the running application with only a few key strokes. Although there are no fundamental barriers against implementing similar features in other language environments, for example, C++, the long history of LISP shows in the quality of the available implementations.

In GINA, the incremental development is further supported by always having a runnable application. There is no need to copy together the skeleton of a program. This skeleton is already provided by the empty application. The programmer simply derives subclasses of GINA classes and overrides the methods as needed. It is not necessary to override all the desired methods at once, because GINA usually provides a default implementation that can already be observed.

GINA provides an inspector facility that can be used to show the slots of arbitrary CLOS objects. References to other objects can be followed by double-clicking on a slot. Objects to be inspected can be found either by starting from one of the main GINA objects (application, document, or main view) or by clicking on an object in the interface (with a special key combination), which shows the underlying CLOS object.

A nontrivial problem with appliction frameworks is how to tell the programmer which subclasses to create and which methods to override to get the desired result. For this purpose, the GINA source code contains a semiformal documentation embedded as comments and made available in different ways.

First, the reference documentation for GINA is automatically generated from the embedded comments. Second, the GINA browser facility allows this information to be accessed online. And third, class definitions and method stubs can be generated by the browser to save some typing.

The documentation facility is implemented by redefining some key CLOS macros such as defclass and defmethod with GINA versions (defginaclass, defginamethod). The GINA versions accept some more arguments, specifying, for example, whether a method may be overridden safely, whether it is called by GINA or by an application, whether there is a default implementation, and so on. All the information that can be extracted by this macro (including the normal CLOS information such as superclasses, default values) is written into a database in the LISP system when the code is loaded.

The GINA browser directly accesses this database. If, during development of GINA itself, changes are made in the code and the code is recompiled, the new information is directly reflected in the browser. In this way, the possibility is reduced that the documentation may become out of date.

For the reference manual, the database contents are read, formatted into TeX statements, and the resulting TeX document is then printed (generating references, index and table of contents).

To speed up typing in methods to override, the browser contains a facility to specify a subclass name and the methods to override, and the browser creates code stubs with the new class name filled in. In this way, typos are avoided and there is no need to cut, paste, and modify such typical code segments.

The Interface Builder

The process of constructing application windows out of individual Motif widgets is supported by a tool, the GINA Interface Builder (Berlage, 1991b). The interface builder allows the application designer to select and parameterize the widgets interactively instead of writing code. Because the designer can quickly check out alternatives, the definition process is accelerated and the final solution is improved.

A number of interface builders have been created—research systems (Cardelli, 1988; Singh et al., 1990) as well as commercial implementations (Webster, 1989). The GINA Interface Builder differs from research systems because it uses a commercial toolkit, and thus provides special support for Motif. It differs from commercial systems using Motif because it provides advanced handling of Motif features such as geometry management and because it integrates its output with the GINA library.

Each application window is designed on a large, scrollable work area (Fig. 5.9). The designer may drag individual widgets from the palette and drop them onto the work area. The widgets on the work area can be freely moved. They are real widgets and not only pictures, so push-button widgets can be pressed, scroll-bars moved, and list entries selected.

The resources of the widgets are specified using modeless dialogs. Every widget has its own dialog window. Changes in the dialog take immediate

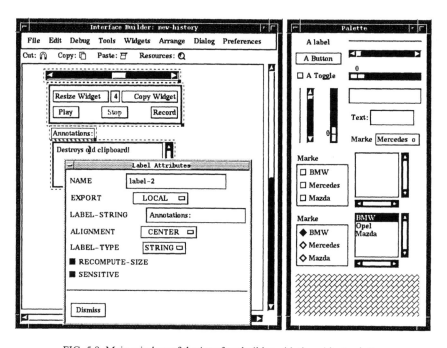

FIG. 5.9 Main window of the interface builder with the widget palette.

effect—entering a single character for the label string of a push-button immediately updates the button widget in the work area that grows to include the new character. Selecting the "Undo" menu entry removes the recently typed characters in both the widget and the dialog and the widget shrinks to its old size.

In Motif, a hierarchy of manager widgets determines the window layout (Fig. 5.10). In this respect Motif differs from many other systems where layout is specified as fixed coordinates or as constraints on a single level. A manager widget completely defines the layout of its components, determined by its layout parameters.

FIG. 5.10 Widget hierarchy for a simple dialog window.

Manager widgets may contain a number of other widgets as components and dynamically adapt sizes and positions of their children, for example, when the user resizes the window. This geometry management process has two goals:

• The widgets negotiate the optimum initial window size with each other. The optimum size depends on end-user preferences such as font sizes or language-dependent labels.

• The user may arbitrarily resize the defined window. A reasonable geometry management ensures that important elements stay visible whereas others may grow or shrink.

To specify the layout, components on the work area are selected and then grouped by one of several possible manager widgets. As soon as widgets are components of a manager widget, they can no longer be independently selected, resized, or moved. This is a consequence of using real manager widgets to exactly observe their behavior.

However, the interface builder employs a grabbing technique to address individual components. Little icons below the menu bar (Fig. 5.9) represent tools that can be dragged to their operand's location. For example, pressing the mouse button over the magnet changes the mouse pointer into a magnet icon until the mouse button is released. On button release, the widget under the magnet is deleted (cut). In a similar fashion, the magnifying glass can be used to pop up the resource dialog for an arbitrary widget, even if it is contained in a manager widget.

By resizing the manager widget, the designer may immediately observe the layout changes that would result when the end user resizes the defined window. The layout parameters can be changed in the resource dialog of the manager widget.

In Motif, the form widget is used for most complex layouts. The geometry management behavior of the form widget is specified by associating an attachment with each side of each component. An attachment is a one-way constraint: the attached side depends on some reference point. The form widget provides eight different types of attachments (e.g., to the side of the form with an offset, to another child with an offset, or to a relative position).

Specifying a layout through attachments is rather laborious. One reason is the amount of detail involved. Furthermore, it is easy to accidentally create useless or contradictory attachment configurations. To avoid these problems, the GINA interface builder uses a less detailed model of rulers that is internally mapped onto the Motif attachment model (Fig. 5.11). The idea is to combine some similar attachments into a ruler (Berlage, 1992a).

The form layout can be changed by direct manipulation in a ruler view. The interface builder translates all changes into Motif attachments. The new layout produced by the form widget is immediately reflected in the ruler view.

The GINA interface builder is able to infer the layout parameters from the component locations before grouping them. For example, the orientation of a row–column widget is set to either horizontal or vertical, depending on whether the selected components were placed more horizontally or more vertically. The same applies to the form widget, where an appropriate ruler configuration is deduced from the approximate component positions established by the designer.

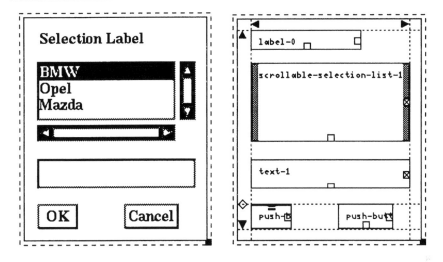

FIG. 5.11 Normal view and ruler view of a form widget.

The interface builder is an application constructed with the GINA class library. According to the GINA document model, the window definition can be saved in a file to be retrieved later.

The result of a window specification is code for GINA that defines a new window class. This class provides a constructor function that can be used to create any number of instances of this class. All components of the window are available as instance variables of the window object to allow dynamic changes of widget parameters.

Conclusion

Our experience has shown that an object-oriented application framework is very well suited to simplify the construction of applications with a graphical user interface. An interface builder is not an alternative but an important supplement for the framework. It is an excellent tool to define the layout of windows, but there is much more about a user interface. Using an application framework, it is also possible to predefine the behavior of the interface. For example, the algorithms and dialogs for opening and closing documents are completely defined in GINA.

An important result of the project is that an object-oriented representation of the interaction history can be used for many other purposes than just a simple undo facility. In particular, we have used the history mechanism to support cooperative work (see chapter 4, section 4.4).

In a system as complex as Motif and X, it is necessary to explore different variations of an application's user interface. The incremental nature of LISP programming, combined with an application framework and an interactive interface builder, makes exploration convenient and fast. On RISC work-stations with enough RAM, the interactive performance of LISP is sufficient. However, delivering stand-alone applications is still a problem due to the size of the LISP run-time system. Where this is important, or where other code must be interfaced, a C++ version of the GINA library can be used.

5.2 PLANNING ASSISTANCE

Joachim Hertzberg

Good clerks think first and then act. This rule may be annulled for routine jobs with well-established procedures for action, where thinking reduces to remembering. But in general, it takes some prior deliberation to do the right thing.

This deliberation comes in different forms, and consequently, different disciplines examine it from different perspectives: Examples are:

- Decision theory, which examines normative models for finding an action that is optimal in some sense.

- Cognitive psychology, which examines human problem solving with the objective of formulating descriptive theories of how humans have come to act as they do.

- Artificial intelligence (AI), which examines computationally effective models of deciding how to act that allow to some degree reasoning from first principles about the domain and about action effects.

The relevant AI branch is commonly called *planning*. Simply put, its main objective is to build systems that, given a representation of a problem, generate a *plan*; executing this plan in the current state of affairs results in another state of affairs, where some *goals* are true that are specified as part of the problem description. We demonstrate later that there are in fact many different instances of planning, but as a least common denominator, one can characterize the term "plan" as denoting "a structure containing representations of actions and goals that serves for reasoning about the effects of future actions, and for influencing their goal-oriented execution" (Hertzberg, 1993, p. 9)

The potential of AI planning for building assistance systems is obvious. In particular, planning can be both a purpose and a means in assistance systems: a purpose whenever the task requiring assistance is a genuine planning task, and a means whenever an assistance task or property requires some internal planning. To take just two instances, consider the assistance properties (Hoschka, 1991) adaptive behavior and explanation ability:

Adaption can be based on plan recognition, as has been done within the AC project (see chapter 2, section 2.2) and outside, as in the WIP project

(Wahlster, 1991); independent of whether the plans to be recognized are hand-coded or automatically generated, using them, as for helping a user find his way out of a faulty system behavior, may often require automatic planning.

Explanation can be and has been viewed as plan based, where the planning task consists in finding a way of expressing the to-be-explained items at the level and modality that fits the user best (which is also an adaptation problem); again, this view has been followed both inside the AC project (see chapter 2, section 2.4) and outside (e.g., Moore & Swartout, 1989).

This text sketches the work on AI planning that we have done within the AC, and the lessons we have learned from this work. Its structure is motivated by the following historical development of the field. Ten years or so ago, planning research focused on building generic planners for what is now called "classical" planning; an explanation is given later. Since then, diversity has grown considerably, adding work and results in the two broad areas of planning theory and of that now known as "nonclassical" planning. Our AI planning work within the AC has followed—and also a little bit influenced— this shift. Consequently, we present it in three sections corresponding to the three fields of classical planning, planning theory, and nonclassical planning.

Note that although this text gives, as a by-product, a brief overview of AI planning, it is not intended to do justice to the work done by researchers or groups other than ours. McDermott (1992) and Hertzberg (1993) give less parochial overviews; (Allen et al., 1990) is a source book including less recent work.

Classical Planning

Early work in planning was heavily influenced by the STRIPS/SHAKEY experiment (Fikes & Nilsson, 1971; Fikes et al., 1972): Given a description of the recent state of the world, a description of desired facts to be made true, and descriptions of executable actions in the world, the STRIPS planner generated a sequence of actions to be executed by the autonomous robot SHAKEY. The overall system also allowed for a certain degree of recovering from plan execution errors and of reusing formerly generated plans.

Motivated by the STRIPS success, follow-up work focused on working within the framework of assumptions and restrictions that the STRIPS designers had made, and dived deep into questions of enhancing planners within it. Only gradually did it become apparent what this framework in fact did consist of, and which STRIPS features were just unimportant design decisions. The framework was then labeled "classical planning" and essentially required that the following items be true:[1]

Domain: States of the world can be completely described by situations (i.e., "snapshots" of what is true); an action applied in a situation yields a situation deterministically; metric time is irrelevant; the information about the domain is complete and accurate at both planning and execution time; all planning goals are explicitly given; there are no hard real-time constraints for planning.

Planners and plans: The planner returns only executable plans that achieve all goals when executed; all planning is from scratch; plan quality does not depend on the time at which they are generated.

Consider as an example the BEAUTIFIER domain as sketched in chapter 3, section 3.3. Classical planning can be used to help beautify a given graphic; in fact, we have made an experiment (Gordon et al., 1993) to solve this problem with a generic classical planner. Remember that the BEAUTIFIER gets as input a draft graphic that may contain certain aesthetical flaws, and proposes a graphic with the flaws fixed. In a prephase, the draft has to be criticized and the flaws listed; the system described by Gordon et al. (1993) uses the identical software as the original BEAUTIFIER for this purpose. As a result, we have an initial situation, namely, the draft graphic, and a set of goals, namely, the to-be-corrected flaws. In the example shown in Fig. 5.12(a), you see a line, let us call it L, and a rectangle, R; L is nearly, but not exactly vertical, and it nearly, but not exactly, touches the center of R's bottom line, which we number as line 1. Let the respective goals for changing these flaws be coded by the two facts:

Vertical(L)
Line-pointing-to-midpoint-of-side (LINE-START-POINT(L)$L,1,R$)

Assuming that the draft graphic is described appropriately, we have a classical planning problem, involving this description and the two facts that are to be achieved by appropriate changes of the graphic.

The latter point is crucial: It is *not* intended to delete the draft and draw a completely new graphic containing two elements L,R with the desired relationships. Instead, the original graphic should be conserved as far as possible,

---

[1] Even today do we have no unanimity about what classical planning is exactly. We here state features that most planning researchers will probably agree on.

and it is assumed that this can be achieved by executing only local changes to single objects.

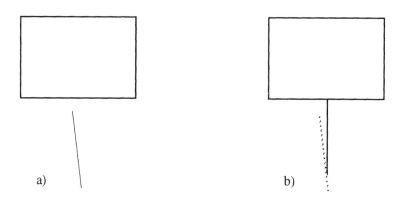

FIG. 5.12 An ugly graphics a), and its beautification b), with the original line dotted.

The third part of a problem description in classical planning, besides the initial situation and the goal predicates, is a set of operators representing actions in the domain. In the graphic domain, we have, for example, operators that represent translating or reshaping objects, bending lines, aligning or making coincident objects, or the like. In classical planning, operators are specified by stating preconditions, that is, predicates that must be true so that the operator is applicable, and postconditions, that is, predicates that will be true whenever the operator has been applied. Following the tradition of STRIPS, we state postconditions in two parts: the added conditions, that is, conditions made true, and the deleted conditions, that is, conditions made false. The situation resulting from applying an applicable operator in some situation, or set of predicates, S, is then S minus the deleted plus the added predicates.

Examples for operators in the graphic domain as used by Gordon et al. (1993) are, first, an operator for making the line l vertical:

BEND-VERTICAL-1(l)
pre: { Free-x(LINE-START-POINT(l)) }
delete: { Horizontal(l) }
add: { Vertical(l) }

and, second, an operator for translating a line *l* such that its start point touches
the midpoint of side *i* of rectangle *r*:

TRANSLATE-LINE-START-TO-SIDE-MIDPOINT-1(l,i,r)
    pre: { Free-x(LINE-START-POINT(l)),
              Free-y(LINE-START-POINT(l)),
              Free-x(LINE-END-POINT(l)),
              Free-y(LINE-END-POINT(l)),
              Free-x(RECTANGLE-POINT1(r)),
              Free-y(RECTANGLE-POINT1(r)),
              Free-x(RECTANGLE-POINT2(r)), ... }
    delete: { Free-x(LINE-START-POINT(l)),
              Free-y(LINE-START-POINT(l)),
              Free-x(LINE-END-POINT(l)),
              Free-y(LINE-END-POINT(l)),
              Free-x(RECTANGLE-POINT1(r)), ... }
    add: { Line-pointing-to-midpoint-of-side(LINE-START-POINT(l),l,i,r)}

Note that representing actions is a tricky issue involving many representation
decisions. The preceding examples mirror a number of them: Points of
objects are initially Free to be moved in *x* or *y* directions, but may be
"frozen" to keep certain predicates true once they are achieved; an operator
making some line Vertical may be "implemented" in different ways, namely,
moving its start point parallel to the *x* axis (thereby shortening the line),
turning it around its end point (thereby changing the *x* and *y* coordinates of its
start point), and others. Whatever possibilities shall be executable for the
system must be properly represented as operators to be considered by the
planner.

A plan to solve our tiny example problem is to apply in this order the operators:

BEND-VERTICAL-1(L)
TRANSLATE-LINE-START-TO-SIDE-MIDPOINT-1$(L , 1, R$)

The other order would not work, according to the operator descriptions,
because translating the line would freeze its border points,which could then
no longer be bent for making the line vertical.

That looks pretty easy, but a "real" representation of the graphic domain
would have to list a tremendous number of different operators, and draft
graphics usually consist of more than just two objects, both facts causing the

search space for a planner to grow to a considerable size. It has in fact been shown (Bylander, 1991) that even much-restricted instances of classical planning problems are PSPACE complete in terms of the problem size, which involves the number of objects and operators in the domain—that is, planning, even if it is decidable, falls into one of the nastiest complexity categories.

Moreover, it can easily be shown (Chapman, 1987) that small and very plausible-looking enhancements to the operator language that still fall into the classical planning framework allow arbitrary Turing machines to be expressed as planning domains, which means that it is in general even undecidable whether there is a plan to solve a problem in classical planning.

Although these results should not deter users or researchers from planning, they teach a certain modesty with regard to what can be achieved easily.

Originally, the central objective of our work in classical planning was to develop a neat conceptual framework for building planners. This framework was to be applied in a toolbox of software modules, the `qwertz`[2] toolbox, intended to assist in building planners. In retrospect, the conceptual work was a success. To give examples, a planning textbook (Hertzberg, 1989), which develops the concepts coherently, has become a standard reference for teaching planning at German-speaking universities; Hertzberg and Horz (1989) concisely described a main subproblem in classical planning: conflict handling. There is a usable version of the `qwertz` toolbox (Gordon et al., 1994), written in the programming language Standard ML.[3]

The toolbox has been used to write a number of planners, such as the one used by Gordon et al. (1993) for the graphic beautification problem. However, the current published version of the toolbox is lacking something that you would probably assume a planning toolbox to include. It contains general artificial intelligence (AI) modules for handling things such as heuristic search and reason maintenance; moreover, there are even more general modules providing implementations of sets and streams and the like, which you may find in other standard programming language libraries (but not in the ML libraries that were available at the time). But there is nearly nothing planning specific in the `qwertz` toolbox, such as signatures for plans or

---

[2] No acronym. The respective GMD project existed from 1988 to 1992.

[3] The `qwertz` toolbox is publicly available by anonymous ftp from the server ftp.gmd.de under GMD/ai-research/Software/qwertz.tar.gz.

planners or operator languages, which leads to one of our most important results in classical planning—albeit a nonfeasibility result.

An important lesson we learned from designing the toolbox is that this original intention was a bit rash. It did not properly take into account how severely planning changes technically whenever you modify planning techniques within the classical framework, let alone extend this framework by just a little bit. For instance, there is just no sensible way to scale up the standard algorithm for classical linear STRIPS style planning (called MZAMZK in Hertzberg, 1989) to a standard algorithm for classical nonlinear planning, although nonlinear is a straightforward generalization of linear planning. The only feasible way by now to integrate the two is to start from the more general algorithms and "cut off" generality, for example, artificially restrict a nonlinear algorithm so as to produce only linear plans. However, that is generally an unsensible approach, because it yields algorithms with the computational overhead of general techniques, but the output of specialized ones.

Obviously, we were not alone in this tangle. For instance, in his view of "standard" planning components, Tate (1994) still left this problem unaddressed. However, this is not just a blind spot in planning theory, but a serious hindrance for building families of practical planners that are supposed to both be usable in application domains of different characteristics and reuse software components as far as possible (Hertzberg, 1994).

In sum, the state of the art in AI planning is not such that it allows generic classical planners to be generalized and to cast this generalization into software tools that are flexibly reusable in classical planners using different techniques: The currently most general possible approach here is still—as in the seventies—building full generic planners. The same is true for the attempt of sharing software tools between nonclassical planners. Consequently, there are no such tools in the qwertz toolbox, leaving this an open project goal, but providing as a result a deep understanding of why this must currently be so.

Planning Theory

When AI planning was in its infancy, planning systems were mostly experimental in character. Boldly put, the working mode was like this: Pick a domain that everybody would agree is a planning domain, write a working system (which would normally turn out to be harder than expected), and describe what you have done. This was the attitude behind the STRIPS work and most of the other planning work until about the mid-eighties—and it is alright: To get a common feeling for the practical issues in planning, this was an effective strategy.

However, it is hard to describe what you have really done when working in this mode. That was the source of Chapman's often-quoted remark about his attemtps to reimplement parts of Sacerdoti's (1977) classical NOAH planner: "Four readings of (Sacerdoti, 1977) and three misconceived implementations later, I had a planner that worked, but no idea why" (Chapman, 1987, p. 334). Good theory can obviously influence and ameliorate practical work. Consequently, many planning researchers have felt the need for a neat theoretical foundation of planning.

The first obvious step toward such a foundation is to develop a common set of notions and concepts in terms of which to formally describe planning work. Its development has in fact started early but usually had some delay, sometimes a tremendous one: For example, essential features of STRIPS, which was built in the late sixties, were not described until 1986 (Lifschitz, 1986). Some of our own work (Hertzberg & Horz, 1989) is of this conceptual character, formally describing notions and techniques dating back to NOAH's times.

However, we think theory has more to offer than just doing the tidy-up work that consists of describing *ex post* what experimentalists have done *ex ante*. A particular problem we have experienced had to do with the attempt to extend modules of the qwertz toolbox to a little more than just the very basic and central notions of classical planning: What happens if some of the just described assumptions of classical planning are relaxed? Which planning notions and techniques will survive such a change unaltered, and which have to be modified, and how?

You will find close to nothing in the literature that answers these questions. On the other hand, they do not seem to be useless or wrongly put. We have argued (Hertzberg, 1993) that knowledge of this kind is essential for developing AI planning from an experimental craft to an engineering science that allows predictable systems to be built.

A prerequisite for answering these questions seems to consist in a vocabulary for characterizing planning domains in terms of their structural properties, sometimes called a planning "ontology." An example for such a characterization was given earlier by the description of classical planning. Two proposals for such a vocabulary are described by Sandewall (1992) ("epistemological specialties") and Hertzberg (1993, section 2); both do not seem completely satisfactory for a number of reasons, but this is not an issue here. Developing a suitable vocabulary is a central problem for planning theory.

We can take an even more extreme view (Hertzberg, 1994). The ultimate form of theoretical planning (as opposed to practical and applied planning) should be axiomatic: State the axioms that a planning application domain is assumed to satisfy, and examine the consequences for planning. Obviously such consequences can be used to develop correctness criteria for plans, to state handy planning procedures that—different from large complex planning systems—can be completely presented and analyzed, or to formally define the calculus underlying the reasoning about action effects. This view is simply a generalization from the experiences made several times while developing the qwertz toolbox: in many cases, obvious, elegant, and seemingly simple generalizations of planning procedures or subprocedures required major research in the axiomatic mode, the result being in many cases that the generalization isn't simple at all, or even possible; again, Horz (1994) presents an example.

The point is that in many cases, it is more efficient to run systematic theoretical studies than to respond to failure when trying to implement a system. Just to mention that we are not alone with this view, complexity analyses as done by Bylander (1991) and Bäckström (1992), Tenenberg's (1991) formulation of planning with precondition elimination, Dean and Boddy's (1988) development of time-dependent planning, and Allen's (1991) well-known work on planning under a rich time model are examples of this working mode.

Our own technical work in theoretical planning has centered around the notion of limited correctness of planners (Hertzberg & Thiébaux, 1994). The idea is to define correctness of a planner by measuring its plans against some action formalism. Such formalisms are the objects of investigation in the field of reasoning about action and change, and there are quite a lot of them, with widely varying expressiveness; Sandewall (1992) gave an overview. However, these formalisms are usually far from being efficiently implementable, and requiring that a plan returned by a planner be strictly correct with respect to some suitable formalism would amount to requiring that the planner be impractical under even the mildest real-time constraints. Limited correctness yields a form of anytime planning (Dean & Boddy, 1988) with the effect that a plan returned gets more correct the more computation time is available for planning. Hertzberg and Thiébaux (1994) explained the central concepts and presented a case study in building such a planner.

Nonclassical Planning

Time-dependendness is a keyword leading to a whole bunch of planning topics that are commonly labeled as "nonclassical". The name is characteristic: In fact, the only point that these topics have in common is that they differ from classical planning in some respect.

Consider again the BEAUTIFIER example. As sketched earlier, it is possible to squeeze this application into the Procrustean framework of classical planning, and maybe the resulting system does even all its users would want it to. However, a number of issues in graphic beautification as such are not covered by the classical framework. For example, a usable assistance system for drawing graphics should respond within relatively tight time bounds; consequently, you cannot afford arbitrary computation times—that is, quick acceptable solutions are better than optimal ones after minutes of computation. Moreover, the goals in beautification are not really definite, and a beautification planner should exhibit the according flexibility: Even if the graphic critic works perfectly well, the user may wish to override certain beautification goals for "semantic" reasons; for example two nearly but not exactly identically high rectangles shall remain as they are because they are to be interpreted as bars in a bar chart, or the ugly graphic in Fig. 5.12(a) shall remain as it is because it is intended to be ugly.

The same applies to many domains: By eliminating features, they could be "classicallified" but the respective planner would be of little use. Consequently, nonclassical planning methods have become a major concern in planning within the last years; McDermott (1992) gave an overview. Our own results here are grouped around the two topics *time* and *uncertainty*, which we sketch in turn.

Temporal Planning

Time here refers to the planning domain, not the planner's computation time. The issue is to deal in planning with metric time as in durations of actions, deadlines, or scheduled events, but also with reasoning about the consequences of, say, overlapping execution of operators, which may alter their effects, compared to the simple sum of the effects of their execution in isolation.

All this requires temporal reasoning abilities on the planner's side that go beyond the simple handling of operator orders. With a job-shop scheduling application in background, we have examined the potential of reusing for planning existing temporal reasoning techniques and technology, namely, *time map management* (Dean & McDermott, 1987).

The idea is simply to have the time map manager (TMM) represent and— well—manage all temporal information like durations of operators or conditions, and to have the planner manage the "logical" dependencies between operators. Implementing this idea requires representing operators and conditions in the TMM in a special way and designing the TMM a little differently from the standard as set by Dean and McDermott (1987), in particular as regards the technique of persistence clipping. But in general, the idea works as simply as it sounds and allows a prototype temporal planner to be built very easily. Rutten and Hertzberg (1993) described how we have done it.

However, one of the problems to be addressed in planning theory is lurking here. One should think (and we *had* thought) that classical planning is just a special case of this form of temporal planning: Briefly, just represent all operators as having 0 (or some tiny real-valued ε) duration. Horz (1994) showed that this does not yield the intended result, with the effect that the "classical" part in our temporal planner architecture, namely, that part dealing with the operator dependencies, must be slightly different from its genuine classical counterpart that is responsible for the same functionality.

Planning Under Uncertainty

Uncertainty is an unavoidable feature whenever building assistance systems (Hoschka, 1991), and it comes in many different forms in planning. For instance, the planner may have inaccurate, incomplete, or dead wrong information about the domain, the world may have changed during planning (e.g., the user has beautified or uglified a graphic himself or herself while the beautifier is running), operator execution may fail, or operators may yield ambiguous effects in principle. Technically, these different forms of uncertainty require being handled by different techniques, so that the label "uncertainty" covers a broad spectrum of work in planning. However, for the abstraction level of this text, it is unimportant to understand in detail technical consequences of these differences. Suffice it to say that only one thing is certain in planning: In general, a plan will not be executed as a planner has designed it to be.

As for the depressing complexity results reported earlier, one reaction to the ever-present uncertainty might be to quit planning altogether, and the advocates of reactive systems (Agre & Chapman, 1987) have done mostly so.

Among those remaining in the planning camp, the reaction is more moderate: Whenever uncertainty is not negligible in some domain, keep planning, but weaken your demands, reinterpret what a plan is and is for, and change the model of how a plan is to be executed. Now it works more like an abstract specification of some default procedure whose execution would normally require additional short-horizon fine-grain planning at execution time to fix the details, and may require updating the whole plan if matters turn out to be grossly different than suspected at planning time. A plan is then no longer a recipe to follow blindly, but a resource for use by a relatively intelligent agent.

We have followed three lines of work to deal with these issues. First, Brewka and Hertzberg (1993) described a formalism and calculus for representing and dealing with incompleteness of or uncertainty about the initial situation of a planning problem, with context dependency of action effects, and with multiple alternative action effects. The idea is—extending the lines of Ginsberg and Smith (1988)—to view a formula as describing a set of possible models one of which describes the "actual" world, but you don't know which. Actions then map possible worlds into possible worlds, where the result of an action is, in general, different for different worlds it is applied in.

Second, building on the results of the previous point, we have developed a plan format and a planner in the possible models formalism (Thiébaux & Hertzberg, 1992). The plans look much like the situated automata or world automata developed by other researchers in planning under uncertainty. The essential idea is that a plan prescribe what to do in each situation. The good thing about all these plan formats is that plan execution is very flexible and able to react to even the most improbable course of events—in many cases even without replanning. The bad thing is that such plans are considerably more complicated to execute than, for example, classical plans, which unconditionally prescribe some course of action. However, if you explicitly assume that things will go wrong, then don't complain if the respective plan execution model asks you to make sure of what is true and what isn't before you continue acting.

On the other hand, many domains allow expectations to be formulated about what will most likely result from executing some action, or which of the many possible initial situations is most probably real; a good strategy both for planning and execution would then be to care about these most likely matters first, and to deal with the unlikely ones only if necessary or if time permits. This was our third line of work (Hertzberg & Thiébaux, 1994); we approached it by blending the possible models formalism with Nilsson's (1986) probabilistic logic. The probability information is a means for defining a component of a realistic measure of plan quality: In general, a plan is better, when it is more complete, that is, the more possible outcomes of actions it tells a reaction to. Using the probability information, one can scale the incompleteness: It does little harm to have no reaction for very unlikely action outcomes. By the way, this is also the central idea behind achieving the limited correctness mentioned in the planning theory section earlier. Hertzberg and Thiébaux (1994) described both the concept and the prototype planner we implemented.

Conclusion

Planning is an indispensible field of methods for building assistance systems in the future. However, the state of the art is not such that you can pick the suitable technology for your favourite application right off the shelf. As sketched in this text, our business within the AC context was to develop planning methodology a bit further into this direction.

Our work has contributed in particular to theoretical planning and to planning methods. To summarize some of the highlights, we have solved a number of open problems, such as integrating a temporal planner and a time map manager (TMM); we have pushed forward the formal understanding of practical planning issues, such as presenting a coherent theory of handling conflicts in classical nonlinear plans; we have shown a new way of defining the central concept of correctness of planners with respect to action calculi, developing the notion of planner correctness in the limit; and we have contributed to the recently starting development of planning "ontologies."

In general, assistance requires planning abilities in domains that are demanding for planning techniques, involving, for example, various facets of uncertainty, cooperation between and coordination of different planners and agents, assessment of the planner's own competence, and timely responses. All these abilities—and most of all their combination—are far from being understood in generality, but isolated solutions exist for specific instances. We are convinced that there will be no breakthrough in putting it all together practically, without a breakthrough in planning theory. On the other hand, there will be no breakthrough in planning theory without running experiments with planning in demanding applications.

Acknowledgments

The heart of the work reported here was done in the context of the `qwertz` project, where the graphic application and the nonclassical work were motivated by our participation in the TASSO project. I thankfully acknowledge the contribution of my colleagues Tom Gordon and Alexander Horz, without whom `qwertz` would never have been the success and the fun that it was.

5.3 METAPHORS AND SYSTEM DESIGN

Peter Mambrey, August Tepper

In this section we report on some findings of a research project[4] in which we investigated the instrumental use of guiding visions and metaphors for technical design and for preventively oriented technology assessment and suggested adequate tools.

This work was part of the institutes' efforts to use the assistance metaphor as guiding vision for system design. From a social scientist's point of view we analyzed the role of this metaphor within different projects to gain deeper information about the daily use in system design and its potentials for technology assessment.

Multiple factors control innovation and development of information systems. The design of such facts and artifacts is a complex process of social construction (Bijker et al., 1993). It is based on societal evolution and individual flashes of insight from inventors. Organizational and societal factors and their local interpretations by individuals determine the development. In addition to other factors, metaphors and guiding visions seem to play an important role in these processes. They attract more and more attention from scientists, analyzing the technological development (Rogers, 1990; Dierkes et al., 1991). Therefore language plays an important role, especially those expressions that are used nonliterally (Carbonell, 1982). These are metaphors, guiding visions, myths, paradigms, and so on. In computer science as a very young science, which includes multidisciplinary approaches from natural sciences, social sciences, and engineering, highly abstract conceptualizations of innovative ideas are widespread (Kemke, 1992). The use of guiding visions and metaphors is therefore ubiquitous; think of the "computer, mouse, desktop" and others. Large differences exist in explaining how metaphors work but very few about the functions. Metaphors and guiding visions must be understood by different persons and must help them to share problems and anticipate future solutions. Metaphors open up perspectives and can therefore

---

[4] This section is based on research done in the project "Metaphors in Computer Science - Potentials and Risks" (in German: *Leitbilder in der Informatik - Potentiale und Risiken. Untersuchungen am Beispiel des Assistenz-Leitbildes*). The project started in October 1990 and ended in September 1993. It has been funded by the German Federal Ministry for Research and Technology.

be used as imaginative mechanisms, similarly to scenarios or other techniques, to prescribe a possible future. They cannot predict or discover utopia but open perspectives to certain aspects, highlighting them, and obscure others, downplaying them (Mambrey & Tepper, 1992; Mambrey et al., 1994).

Paradigms, *Leitbilder*, and Metaphors

A basic problem of using visions and metaphors for technology development and technology assessment deals with the meaning of cognitive processes and conceptual ideas in nature. It would be relatively easy if our concepts—including visions and metaphors—might directly represent the ontic world existing independently of our consciousness. Then a vision would not simply be an idea, a cognitive possibility, but contain ontic propositions and would be a directly useful source for any designer (Danesi, 1990). However, such an immediate connection is not available, but visions and metaphors are human constructions—as any other concept—that are useful to structure worlds and that are therefore used (Helm, 1992; Way, 1991). This is sufficient as a basis here, and frequently made assumptions about ontic qualities of pictures and so on are not necessary and also incorrect.

As a tool for analyzing and designing such constructions, three basic concepts are needed: paradigms, guiding visions (*Leitbilder*), and metaphors. Paradigms and guiding visions resemble each other only at first sight, they should be clearly differentiated. According to Thomas S. Kuhn (1962), paradigms are long-term orientation patterns of science, such as the classical Newtonian mechanics, that were for long periods unquestioned and were reused and proved again and again by "normal" science, as Kuhn calls it. These are the basic selections in a complex world that decide on whether concepts such as "guiding visions" are considered relevant at all.[5] Here it is only necessary to remember the concept of the paradigm as necessary background of a future assessment by metaphors that cannot be questioned directly for lack of an observation position situated at a higher level.

---

[5] Thomas S. Kuhn demonstrated this fact with a plain but momentous statement about the very basic requirements of research. Even the question of what the researcher intends to study requires a previously given picture: "The commitments that govern normal science specify not only what sorts of entities the universe does contain, but also, by implication, those that it does not" (1962, p. 7). According to Kuhn, no natural science is possible without an at least implicit complex of interconnected theoretical and methodological convictions making selection, evaluation, and critique possible. Without such models, research activities could not be started at all, because otherwise all facts to be observed would be of identical relevance.

Guiding visions and metaphors are like the two sides of a coin. For illustrating different aspects and for the conceptual separation thus required, guiding visions (*Leitbilder*) denote the function and metaphors denote a frequent and important form of expression. The function of guiding visions is that of a medium[6] that establishes "structural interconnections" between different and largely autonomous social subsystems. By analogy with Talcott Parsons, Niklas Luhmann (1984) called them symbolically generalized communication media, and a very simplified example is public opinion. Public opinion results from an interaction between journalism and politics. Both sides supply news and commentaries, and a great number of only loosely coupled opinions are replaced by a construct to which politics and media strictly adhere. The symbolism that is assigned to the construct is important. Because of their advantages, media such as public opinion or guiding visions are permanently reproduced by the social system as a social instance, namely, by changing and advancing the respective subjects. Dynamics, the change of subjects, is required for the existence of the social instance.

In the case of the assistance metaphor, all functions could clearly be seen. For example, some research projects within the institute where this guiding vision was active accepted the metaphor as given but did not share the perspectives. But soon they recognized that their status and resources were depending more and more on the particular contributions to this guiding vision. So they reoriented themselves and reshaped their work. The question of whether the guiding vision of the assistance computer was obligatory for all research projects, was heavily discussed with no final results. On the other hand, in the de facto daily work, compromises and reorientation towards the new guiding visison took place step by step.

If a *Leitbild* is expressed already in more or less precise terms, these descriptions can easily be used for discussing technological improvements on an early phase of design to gather orientation. But *Leitbilder* are relatively seldom expressed in precise terms, especially in the early stages of a new technology. Quite often visions cannot be explained by using existing words with a fixed meaning, and people are forced to compose new terms like "paperless office", "knowledge navigator", or "assistance computer." A metaphor is one

---

[6] In sociological theory, the term "medium" is not used in the same sense as in computer science or communication engineering. The term was created in response to the problems of a specific theory construction dealing above all with the separation of society into functional subsystems and is now confronted with the question of how to enable integration (or in more general terms, social order).

of the means to produce an utterance of a vision. Therefore the analysis of metaphors used by researchers is not only necessary but much more interesting.

Metaphors are forms of a nonliteral use of language. Metaphors interactively combine known and new aspects; they explain new aspects in terms of similar ones. The structural and conceptual force of the metaphors lies in their surreal quality and polysemantic. They indicate a course without tracing it out in detail. It is important, that metaphors are always linguistic entities and not pictures. A picture would be an object about which one can speak with the aid of metaphors, for example.

Metaphors have a lot of functions, but expressing creative thinking, directing work, allowing communication about an object that does not yet exist in reality, and making explanations easier are the most important ones. To some extent, metaphors thus secure user orientation automatically, because the respective systems are designed within a structure that is more or less familiar to future users.

Assumptions About How Metaphors Work

In addition to the question of what metaphors are, it is interesting how they work. This question is answered differently. Today Max Black's interaction theory is widely accepted as a general model of how metaphors work (Indurkhya, 1992), but older and alternative theories might explain at least some special cases.

The old Greek and Roman philosophers thought that words showed something of the real world (substitution theory) and regarded metaphors as a substitute quite often used only for rhetoric purposes. The task was to discover the "true meaning" behind metaphors.

Today the substitution theory is replaced by more complex theories. Some researchers see metaphors as a comparison transfering meanings from the source to the target domain; for example, an atom is structured like the solar system. Objections based on obvious mismatches are rejected by restricting the comparison to salient aspects only (ships and planes have captains in common but not falls or sheets). Extracting these aspects allows statements to be made about certain properties of the target domain. Like other authors, Gentner (1982, p. 108; 1983) restricted the comparison to certain salient aspects. The decisive feature of Gentner's approach is the fact that only the relational structure of the source is mapped to the target, not the properties of the relevant objects. In the metaphor "The atom is like the solar system" the relation "revolves around" between sun and planets is true for nucleus and electron too, but the attribute "the sun is yellow" does not carry over. This model is interesting from a methodological viewpoint because the selection of the salient aspects is supported by a specifying category (even if it is unlikely that all basically possible relations of the source are mapped to the target).

Approaches to metaphor research that use precultural explanations pursue a quite different method of problem solving. Examples are Jung, Levi-Strauss, and Lakoff. According to Jung (1976), archetypes are innate psychological structures, ancient patterns of human behavior. Claude Levi-Strauss discovered in metaphors a metacode that is superexistent for an individual. Lakoff and Johnson pursued an approach that was comparable to some extent (Lakoff, 1987; Lakoff & Johnson, 1980). They traced back language to the use and combination of basic components (semantic primitives). These primitives correspond to very original body experiences such as every thing being either inside or outside of a body, and Lakoff and Johnson outlined a metaphorical process combining elementary concepts like components to more complex utterances: "most concepts are partially understood in terms of other concepts" (Lakoff & Johnson, 1980, p. 56). Thus, according to their hypothesis, metaphorical utterances can be "understood" in a relatively simple way by machines and they can be associated with meanings.

Black's interaction theory is most popular. Interaction theory is very much connected to modern constructive philosophy and does not suppose a substitution or a comparison. Black and other researchers see metaphors as the product of an interaction between source and target producing something unknown before. We do not discover hidden similarities but we construct similarities between different objects by stating and testing such connections. Once you have noticed a new possibility you have mentally created a new object (to which people quite often refer with the help of composed words like "assistance computer"). Metaphors therefore make use of similarities but really name a distinctive new object. Based on this theory, all expressions have literal meanings (either known or new objects), and metaphors can be analyzed in almost the same way as "normal" notions. By interaction, Black (1979) outlined a process that is clearly distinguished from the transference of specific comparison aspects. However, the new interpretation of the metaphorical process still comprises structural analogies as a basis but leads to a new understanding, the projection.

Metaphors as Instruments

Metaphors are used for different purposes. Our idea was to use them explicitly in a structured way as instruments for orientation and evaluation in technical development. Using metaphors instrumentally is not just playing with words; it is playing with visions, possible targets, future aims, and future solutions. It provides rich information for insight and yields a wider range of perspectives. The instrumental use of metaphors therefore can be to structure and enhance joint brainstorming or self-reflection (Mambrey, 1995). The following examples show some structured ways to work explicitly with metaphors, gaining information and orientation in design. They can be used as tools for anticipation.

Exploration by Metaphors

The first step in varying words and expressions is to select a domain and a word or metaphor to analyze. One can think of some guidelines and instruments to support this step, but in the end it comes down to exploration (see Fig. 5.13). Such an exploration also can be viewed as a biased process of cognitive construction. At least once an "arena" is named, the analyst/ designer will find some elements on stage to start with.

At the left side of Fig. 5.13, "mother" metaphors provide the starting point for selecting an analysis metaphor and two examples of metaphor combinations. At the right side of the figure, in the case of a complementary metaphor a representation module, such as a trash icon, is connected to only one functional module. In the case of a multiple metaphor, different metaphors and their representing icons activate the same functional module.

In this context, the idea of inheritance as complementary process to the definition of semantic primitives fits perfectly. The properties of a semantic primitive also mold a complex object. Anyone who has observed the trouble users have recognizing and following processes of inheritance must find the opportunities offered by the new system to illustrate (inherited) similarities quite interesting. The interaction theory is focused on the creative construction of metaphors. The creations may have been produced systematically (e.g. with the aid of known creativity techniques), although it is inexpedient to attempt to retrace the chain of ideas. The newly created metaphorical notion is paramount here—the sense and structure of this construction is a matter of discursive interpretations. This, too, can be graphically illustrated, and would draw the user's attention to the fact that a creative construction is

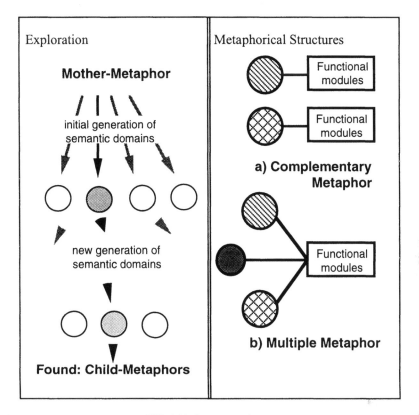

FIG. 5.13 Instrumental usage.

expected from him or her in the part of the system with which he or she is dealing.

Structuring by Metaphors

Normally, research projects have quite complex goals that cannot be covered by just one metaphor. To that end they explicitly or accidentally compose a metaphorical scenario. Sometimes these metaphors are not connected, but quite often a structure can be discovered and applied. Metaphors may be contradictory, or metaphors may be instances of a central metaphor. For example, the Japanese project FRIEND21 developed such structures (Nonogaki & Ueda, 1991). The Japanese "metaphor engineers" aim to build a system out of calculated and combined metaphors (see right-hand side of Fig. 5.13).

They speak about "complementary metaphors" (one metaphor serves one functional module), "multiple metaphors" (more than one metaphor serves one functional module), and "composite metaphors" (one or more metonymies bridge a gap between contradictory metaphors).

Classification by Metaphors

Categorizing a given metaphor is also a promising method. Philology delivers an elaborated system of definitions to distinguish between different kinds of metaphors: These are allegories (figurative sentence whose attributes refer to the subject they are meant to suggest), animation (anthropomorphization or personification as special cases), deanimation, artificialization, compression, simile (comparison of one thing with another), condensation, context overlapping, deformation, functionalization, incongruence, indirect speech ("can you pass the salt?"), inversion, irony (the intended meaning is the opposite of that expressed by the words used), litotes (understatement in which an affirmative is expressed by the negative of the contrary as in "not a bad singer"), metonymy (change of name: "Washington made a statement" instead of "The president made a statement"), overstatement/understatement, oxymoron (cruel kindness, a combination of contradictory or incongruous words: sweet and sour), pars pro toto (part for whole: "young and old" instead of "all"), reduction to subaspect, shaping (outer form of something by which it can be seen or felt to be different from something else), simplification, synecdoche (a more comprehensive term is used for a less comprehensive and vice versa: "all hands on deck"), and transfusion (photographic-pictorial overlapping). Some of these metaphorical categories give important information about its contents and can support assessment quite a lot (e.g., overstatement or irony). Others are used to construct bridges in a metaphorical scenario. A metonymy like the "trash" links the two domains of deleting a file and ejecting a disk, at least in some computer systems.

Ingendahl (1973) examined the metaphorical use of words for semantic domains and semantic features. According to Ingendahl, a metaphor is understood by the relation to the content neighbors within the sense domain (paradigmatic relation), to the neighbors in the context (syntagmatic relation). As a metaphor the word originates from such relations and enters again into such relations; within the metaphorical process, type and reference points of the relation are changed (Ingendahl, 1973, p. 67). These two basic assumptions can be used to reconstruct step by step the meaning of a metaphor and, at least as first aid, this can be supported by programs using structured dictionaries.

Discovering Primitives

An interesting approach oriented to the works by Lakoff and Johnson about semantic primitives is the discovery of basic building components in metaphors. Semantic primitives such as container, war and others are not very useful for future assessment because impacts are hard to derive from them. However, the elaboration of specific primitives for specific technological areas is however most promising. For example, basic ideas for discussing technological impacts can easily be derived from a system component such as "centralization." Each research domain discusses a set of such basic primitives, such as machine or tool. As another example, computer science specifies as special primitives categories like game, society, family, and journey (Kendall & Kendall, 1992). Quite often these primitives have fixed meanings (centralization, for example, is viewed as a bad thing). The task is to find further and/or better primitives and to find the correct expressions for them.

Contrasting

A very similar method is the description and contrasting of different metaphors. Contrasting manifests correspondences between the compared metaphors and shows above all which orientations are different. The comparision shows which—possibly substantial—questions are disregarded by the different metaphors. The confrontation of one's own guiding vision with other guiding visions is an approach pursued by many engineers and others quite naturally. Even if it is only a rudimentary systematic contrasting, it shows that it is no instrument invented exclusively for technology assessment. The developers observe not only the performance of their guiding vision but also their boundaries. However, the found differences often lead to consistent changes of research design in follow-up projects only. Even if we consider an incubation time for new ideas, this process might be accelerated considerably by a systematic analysis of guiding visions. By analyzing the discourse and by contrasting the selected metaphors with possible others one can see important things excluded from the design process. This perspective allows a more extensive impression of competing Leitbilder and metaphors. By comparing and contrasting the metaphors and Leitbilder used in a specific context, the interaction between understanding and creation became observable.

Because every evaluation needs knowledge about the concrete application context, the task of scientific analysis, sketched out earlier, is as limited as any prophecy. But improved discourses, based on described and structured observations, make it possible to discuss solutions.

Simple Example

A simple example of an evaluation process is outlined in the following illustration. Within the framework of our research project "assistance computer," the first part of this compositum was elaborated with the aid of a thesaurus. In addition to encyclopediae and electronic dictionaries, computer systems able of interpreting metaphors may also be useful (cf. Carbonell, 1982; Martin, 1990). A "true" statement on the word domain will not be attained in any instance; instead the selection and arrangement of the synonyms each constitute special cognitive constructs. For the sake of simplicity, Figure 5.14 elaborates on those synonyms shown in bold type (in principle, however, every synonym could be elaborated further). The project workers were first asked to select the terms that were important for them and were then requested to evaluate these.

Figure 5.14 shows the elaboration of the "assistance computer" metaphor into synonyms and evaluation of the latter by a group of developers. A majority agreed that blackened words mark favorable properties of an assistance computer and framed words indicate unwanted qualities.

The encyclopedic search for synonyms expands the semantic word domain of a metaphor. As early as the first stage, the verbs "assist" and "aid" provide a number of important alternatives, which, over the course of the following stages, provide clear information on useful and unsuitable properties of an assistance computer (cf. black and framed areas in Fig. 5.14). The word "assistance" can be seen to embrace only a few useful properties (the word after all derives from quite a different sphere, namely the latin *stare*, stand). The most fruitful results, on the other hand, are obtained from the elaboration of the word "execute." Forms of support of a more psychological nature, such as "encourage" and "invigorate," are clearly shown to be inappropriate properties of an assistance computer.

One result of this simple description and evaluation process is the fact that a clear demarcation line exists between the properties of human assistance and the desired properties of an assistance computer. The general result is a refined definition of the "assistance" metaphor whose contents have in part been redefined and described with greater precision. Such definition refinements are absorbed directly into the development process.

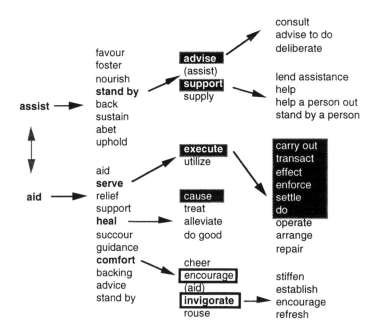

FIG. 5.14 Elaboration of the "assistance computer" metaphor.

Conclusion

A possible application of these instruments is to support researchers and developers in reflecting on the guiding visions and metaphors they use and revealing excluded aspects. They can be used as tools to reduce complexity and as an orientation frame. They make the discussions about new technical systems more explicit and facilitate a discussive consensus. There is no shortcut to utopia. But there are means to construct future technical and social systems that offer a certain amount of control with hindsight over the process. The methodological use of guiding visions and metaphors changes the situation with respect to the following aspects:

Knowledge: Guiding visions and metaphors focus technology-oriented com-munication processes. The latent meaningful relations inherent in the respective concepts can be revealed, changed, and questioned for possible consequences in the case of their realization. It is, however, a special type of knowledge; it is both comprehensive and polyvalent depending on the specific development phase. Polyvalence is step-by-step reduced by the context. This means that a given selection is consolidated, thus becoming trend-setting.

Prognosis: Neither traditional technology assessment nor technology assessment based on guiding visions and metaphors can predict consequences in the strict sense. Therefore, this approach will not change the basic problem.

Control: Guiding visions and metaphors are part of technical development. Any form of more systematic analysis, construction, and evaluation of guiding visions and metaphors is therefore an additional instrument for controlling technical development. Changes to guiding visions and metaphors will change development.

Evaluation: Any evaluation first presupposes an object that can be described more or less precisely. In this respect, evaluation of guiding visions and metaphors is dependent on knowledge. However, this is only one component of the evaluation process, because evaluation is always a very subjective action and includes power and politics. The question is whether the realization of the outlined vision is personally desirable or not. This can be done by verbal dissection.

Expertise is useful for understanding visions and metaphors. However, they are in general formulated so graphically that the relationship to one's own interest is relatively easy to establish. The use of guiding visions and metaphors could also facilitate finding compromises and social consensus. The analysis and discussion of metaphors and guiding visions is thus a participative instrument.

Anticipating the "future" based on guiding visions and metaphors cannot and should not replace other procedures but complement them. Visions and metaphors naturally have the greatest effectiveness in the definition phase of new technologies (Mambrey, 1994) and despite all the advantages and disadvantages, no alternative basis is available for this phase of technology design and technology assessment. Future assessment by metaphors (Tepper, 1993) can be used both as an instrument of self-reflection of researchers and developers, thus avoiding the delicate question of access to the laboratories, and as an instrument of public discussion (e.g., organization of consensus conferences about guiding visions and metaphors). In later phases there are alternatives, but aspects such as the capability of visions and metaphors to condense information might advocate their use.

References

[Ackermann & Ulich, 1987] D. Ackermann and E. Ulich. The Chances of Individualization in Human–Computer Interaction and its Consequences. In: M. Frese, E. Ulich, and W. Dzida, editors, Psychological Issues of Human–Computer Interaction in the Work Place, pp. 131–145. Elsevier Science Publishers, Amsterdam, 1987.

[Agre & Chapman, 1987] P. Agre and D. Chapman. Pengi: An Implementation of a Theory of Action. In: Proc. AAAI-87, pp. 268–272 (Seattle, WA, July 13–17, 1987). Morgan Kaufmann, San Mateo, CA, 1987.

[Alexy, 1989] R. Alexy. A Theory of Legal Argumentation. Clarendon Press, Oxford, UK, 1989.

[Allen, 1983] J. Allen. Maintaining Knowledge about Temporal Intervals. Communications of the ACM 26, pp. 832–843, 1983.

[Allen, 1991] J. Allen. Temporal Reasoning and Planning. In: J. Allen, H. Kautz, R. Pelavin, and J. Tenenberg, editors, Reasoning About Plans, chapter 1, pp. 1–68. Morgan Kaufmann, San Mateo, CA, 1991.

[Allen et al., 1990] J. Allen, J. Hendler, and A. Tate, editors. Readings in Planning. Morgan Kaufmann, San Mateo, CA, 1990.

[Archer et al., 1984] J. E. Archer, R. Conway, and F. B. Schneider. User Recovery and Reversal in Interactive Systems. ACM Transactions on Programming Languages and Systems 6(1), pp. 1–19, 1984.

[Arkin et al., 1989] E. M. Arkin, L. P. Chew, D. P. Huttenlocher, K. Kedem, and J. S. B. Mitchell. An Efficiently Computable Metric for Comparing Polygonal Shapes (Technical Report No. TR89–1007). Cornell University, Ithaca, NY, 1989.

[Ashlar Vellum, 1992] Vellum Software GmbH. Ashlar Vellum Benutzer- und Referenzhandbuch. Miltenberg, 1992.

[Bäcker, 1992] A. Bäcker. Designing Reuseable Widget Classes with C++ and OSF/Motif. The X Resource (2), pp. 106–130, O'Reilly and Associates, Sebastopol, CA, 1992.

[Bäcker et al., 1992] A. Bäcker, C. Beilken, T. Berlage, A. Genau, and M. Spenke. CLM: A Language Binding for Common Lisp and OSF/Motif, User Guide and Reference Manual. Arbeitspapiere der GMD 612, GMD, Sankt Augustin, 1992.

[Bäckström, 1992] C. Bäckström. Computational Complexity of Reasoning About Plans. Ph.D. thesis, Linköping Studies in Science and Technology, Dissertation No. 281, Dept. of Computer and Information Science, Linköping University, Linköping, Sweden, 1992.

[Baecker & Small, 1990] R. Baecker and I. Small. Animation at the Interface. In: B. Laurel, editor, The Art of Human–Computer Interface Design, Addison-Wesley, Reading, MA, 1990.

[Barber et al., 1993] R. Barber, W. Equitz, C. Faloutsos, M. Flickner, W. Niblack, D. Petkovic, and P. Yanker. Query by Content for Large On-Line Image Collections (Technical Report No. RJ 9408 (82600)). IBM Research Division, San Jose, CA, 1993.

[Bartsch-Spörl et al., 1991] B. Bartsch-Spörl, B. Bredeweg, C. Coulon, U. Drouven, F. van Harmelen, W. Karbach, M. Reinders, E. Vinkhuyzen, and A. Voß. *Studies and Experiments with Reflective Problem Solvers.* ESPRIT Basic Research Action P3178 REFLECT, Report IR.3.1,2 RFL/BSR-UvA/II.2/1, REFLECT Consortium, August 1991.

[Bechtel & Abrahamsen, 1991] W. Bechtel and A. Abrahamsen. *Connectionism and the Mind.* Basil Blackwell, Cambridge, MA, 1991.

[Becker, 1993a] A. Becker. Automatisch akquirierte Schwierigkeitsebenen als Basis für Benutzermodellierung. In: A. Kobsa and W. Pohl, editors, *Arbeitspapiere des KI-93 Workhops "Adaptivität und Benutzermodellierung in interaktiven Softwaresystemen"* (Berlin, 13.–15.9.1993). WIS-Memo Nr. 7, AG Wissensbasierte Informationssysteme, Informationswissenschaft Universität Konstanz, September 1993.

[Becker, 1993b] A. Becker. *Ein generisches Benutzermodellierungssystem zur Akquisition und nicht-monotonen Aktualisierung individueller Benutzermodelle.* DIAMOD Report 26, GMD, Sankt Augustin, April 1993.

[Beimel et al., 1992] J. Beimel, J. Hüttner, and H. Wandke. Kenntnisse von Programmierern auf dem Gebiet der Software-Ergonomie: Stand und Möglichkeiten zur Verbesserung. Unpublished paper of a lecture on the *Fachtagung der Sektion Arbeits-, Betriebs- und Organisationspsychologie des Berufsverbandes Deutscher Psychologen "Arbeits-, Betriebs- und Organisationspsychologie vor Ort".* (May 25–27, 1992, Bad Lauterbach).

[Benyon & Murray, 1988] D. Benyon and D. Murray. Experience with Adaptive Interfaces. *The Computer Journal 31*(5), pp. 465–473, 1988.

[Benyon et al., 1990] D. Benyon, D. Murray, and F. Jennings. An Adaptive System Developer's Toolkit. In: D. Diaper, D. Gilmore, G. Cockton, and B. Shackel, editors, *Human–Computer Interaction—INTERACT'90,* pp. 573–577. North Holland, New York, 1990.

[Berlage, 1991a] T. Berlage. *OSF/Motif: Concepts and Programming.* Addison-Wesley, Wokingham, UK, 1991.

[Berlage, 1991b] T. Berlage. The GINA Interface Builder. In: *Proceedings of the First Annual Motif Users Conference,* pp. 184–193 (Washington, DC, Dec. 8–10, 1991).

[Berlage, 1992a] T. Berlage. A Ruler Model for the Motif Form Widget. *The X Resource* (3), pp. 137–153, O'Reilly and Associates, Sebastopol, CA, 1992.

[Berlage, 1992b] T. Berlage. Using Taps to Separate the Application Code from the User Interface. In: *Proceedings of the ACM Symposium on User Interface Software and Technology,* pp. 191–198 (Monterey, CA, Nov. 15–18, 1992). ACM Press, New York, 1992.

[Berlage, 1993a] T. Berlage, editor. *Object-Oriented Application Frameworks for Graphical User Interfaces.* Berichte der GMD 213, Oldenbourg, München, 1993.

[Berlage, 1993b] T. Berlage. *Recovery in Graphical User Interfaces Using Command Objects.* Arbeitspapiere der GMD 774, GMD, Sankt Augustin, 1993.

[Berlage & Genau, 1993] T. Berlage and A. Genau. A Framework for Shared Applications with Replicated Architecture. In: *Proceedings of the ACM Symposium on User Interface Software and Technology,* pp. 249–257 (Atlanta, GA, Nov. 3–5, 1993). ACM Press, New York, 1993.

[Berlage & Spenke, 1992] T. Berlage and M. Spenke. The GINA Interaction Recorder. In: J. A. Larson and C. Unger, editors, *Proceedings of the IFIP TC2\WG2.7 Working Conference on Engineering for Human-Computer Interaction,* pp. 69-78 (Ellivuori, Finland, Aug. 10–14, 1992). Elsevier Science Publishers, Amsterdam, 1992.

[Bertin, 1983] J. Bertin. *Semiology of Graphics*. University of Wisconsin Press, Madison, WI, 1983.

[Bier, 1989] E. A. Bier. *Snap Dragging: Interactive Geometric Design in Two and Three Dimensions*. Ph.D. thesis, University of California, Berkeley, CA, 1989.

[Bijker et al., 1993] W. E. Bijker, T. B. Hughes, and T. Pinch, editors. *The Social Construction of Technological Systems. New Directions in the Sociology and History of Technology*. MIT Press, Cambridge, MA, fourth printing, 1993.

[Birman et al., 1991] K. Birman, A. Schiper, and P. Stephenson. Light Weight Causal and Atomic Group Multicast. *ACM Transactions on Computer Systems 9*(3), pp. 272–314, 1991.

[Bischof, 1989] N. Bischof. Ordnung und Organisation als heuristische Prinzipien des reduktiven Denkens. In: H. Meier, editor, *Die Herausforderung der Evolutionsbiologie*, pp. 79–127. Piper, München, 1989.

[Bisson, 1992] G. Bisson. Conceptual Clustering in a First Order Logic Representation. In: B. Neumann, editor, *Proc. ECAI-92*, pp. 558–462 (Vienna, Austria, Aug. 3–7, 1992). Wiley, Chichester, 1992.

[Black, 1979] M. Black. More About Metaphor. In: A. Ortony, editor, *Metaphor and Thought*, pp. 19–43. Cambridge University Press, Cambridge, 1979.

[Blair & Subrahmanian, 1989] H. A. Blair and V. S. Subrahmanian. Paraconsistent Logic Programming. *Theoretical Computer Science 68*, pp. 135–154, 1989.

[Bolz, 1990] D. Bolz. *Eine Sprache zur Beschreibung graphischer Situationen*. TASSO Report 13, GMD, Sankt Augustin, 1990.

[Bolz, 1993a] D. Bolz. Some Aspects of the User Interface of a Knowledge-Based Beautifier for Drawings. In: W. D. Gray, W. E. Hefley, and D. Murray, editors, *Proceedings of 1993 International Workshop on Intelligent User Interfaces*. ACM Press, New York, 1993.

[Bolz, 1993b] D. Bolz. *Maschinelle Assistenz bei der Korrektur von Zeichnungen*. Dissertation, Universität Bremen, Bremen, 1993. Also as: Berichte der GMD 228, Oldenbourg Verlag, München, 1994.

[Borning, 1979] A. Borning. *ThingLab—A Constraint-Oriented Simulation Laboratory*. Ph.D. thesis, Stanford University, Stanford, CA, 1979.

[Borning, 1986] A. Borning. Constraint-Based Tools for Building User Interfaces. *ACM Transactions on Graphics 5*(4), pp. 345–375, 1986.

[Borning & Travers, 1991] A. Borning and M. Travers. Two Approaches to Casual Interaction over Computer and Video Networks. In: *Proceedings of the CHI Conference on Human Factors in Computing Systems*, pp. 13–19 (New Orleans, LA, April 28–May 2, 1991). Addison-Wesley, Reading, MA, 1991.

[Borning et al., 1987] A. Borning, R. Duisberg, B. N. Freeman-Benson, A. Kramer, and M. Woolf. Constraint Hierarchies. In: N. Meyrowitz, editor, *Proc. OOPSLA '87*, pp. 46–60 (Orlando, FL, Oct. 4–8, 1987). ACM Press, New York, 1987.

[Bowers & Churcher, 1988] J. Bowers and J. Churcher. Local and Global Structuring of Computer Mediated Communication: Developing Linguistic Perspectives on CSCW in COSMOS. In: *CSCW'88, Proceedings of the Conference on Computer-Supported Cooperative Work*, pp. 125–139 (Portland, OR, Sept. 26–29, 1988). ACM Press, New York, 1988.

[Breuker & van de Velde, 1994] J. Breuker and W. van de Velde. *The Common KADS Library: Re-Usable Components for Artificial Problem Solving*. IOS Press, Amsterdam, Tokyo, 1994.

[Breuker & Wielinga, 1989] J. A. Breuker and B. J. Wielinga. Model Driven Knowledge Acquisition. In: P. Guida and G. Tasso, editors, *Topics in the Design of Expert Systems*, pp. 265–296. North Holland, Amsterdam, 1989.

[Brewka, 1991] G. Brewka. *Nonmonotonic Reasoning—Logical Foundations of Commonsense.* Cambridge University Press, Cambridge, 1991.

[Brewka & Hertzberg, 1993] G. Brewka and J. Hertzberg. How to do Things with Worlds: On Formalizing Actions and Plans. *Journal of Logic and Computation 3*(5), pp. 517–532, 1993.

[Brown, 1984] A. L. Brown. Metakognition, Handlungskontrolle, Selbststeuerung und andere, noch geheimnisvollere Mechanismen. In: F. Weinert and R. Kluwe, editors, *Metakognition, Motivation und Lernen*, pp. 60–108. Kohlhammer, Stuttgart, 1984.

[Browne et al., 1990] D. Browne, P. Totterdell, and M. Norman. *Adaptive User Interfaces.* Academic Press, London, 1990.

[Bylander, 1991] T. Bylander. Complexity Results for Planning. In: *Proc. IJCAI-91*, vol. 1, pp. 274–279 (Darling Harbour, Sydney, Australia, Aug. 24–30, 1991). Morgan Kaufmann, San Mateo, CA, 1991.

[Carbonell, 1982] J. G. Carbonell. Metaphor. An Inescapable Phenomenon in Natural Language Comprehension. Metaphor as a Key to Extensible Semantic Analysis. In: W. G. Lehnert and M. H. Ringle, editors, *Strategies for Natural Language Processing*, pp. 415–434. Lawrence Erlbaum Associates, Hillsdale, NJ, 1982.

[Cardelli, 1988] L. Cardelli. Building User Interfaces by Direct Manipulation. In: *Proceedings of the ACM SIGGRAPH Symposium on User Interface Software*, pp. 152–166 (Banff, Alberta, Canada, Oct. 1988). ACM Press, New York, 1988.

[Chandrasekaran, 1988] B. Chandrasekaran. Generic Tasks as Building Blocks for Knowledge-Based Systems: The Diagnosis and Routine Design Examples. *Knowledge Engineering Review 3*(3), pp. 183–210, 1988.

[Chang et al., 1987] S. K. Chang, Q. Y. Shi, and C. W. Yan. Iconic Indexing by 2-D Strings. *IEEE Transactions of Pattern Analysis and Machine Intelligence 9*(3), pp. 413–428, 1987.

[Chapman, 1987] D. Chapman. Planning for Conjunctive Goals. *Journal of Artificial Intelligence 32*, pp. 333–377, 1987.

[Chappel & Wilson, 1991] H. R. Chappel and M. D. Wilson. *A Multi-Modal Interface for Man-Machine Interaction with Knowledge-Based Systems.* SERC Rutherford Appleton Laboratory, UK, June 1991.

[Clocksin & Mellish, 1984] W. Clocksin and C. S. Mellish. *Programming in Prolog.* Springer-Verlag, Berlin, 1984.

[Conklin & Begeman, 1988] J. Conklin and M. Begeman. gIBIS: A Hypertext Tool for Exploratory Policy Discussion. *ACM Transactions on Office Information Systems 6*(4), pp. 303–331, 1988.

[Console et al., 1991] L. Console, D. T. Dupre, and P. Torasso. On the Relationship Between Abduction and Deduction. *Journal of Logic Computation 1*(5), pp. 661–690, 1991.

[Coulon et al., 1992] C. H. Coulon, F. van Harmelen, W. Karbach, and A. Voß. Controlling Generate & Test in Any Time. In: *Proc. GWAI-92*, pp. 110–121. Springer Lecture Notes on AI, vol. 671, Berlin, 1993.

[Danesi, 1990] M. Danesi. Thinking is Seeing: Visual Metaphors and the Nature of Abstract Thought. In: *Semiotica 80*(3/4), p. 227, 1990.

[Davis, 1976] R. Davis. *Applications of Meta-Knowledge to the Construction, Maintenance, and Use of Large Knowledge-Bases.* AI memo 283, Stanford University, Palo Alto, CA, July 1976.

[Davis & Buchanan, 1977] R. Davis and B. G. Buchanan. Meta-Level Knowledge: Overview and Applications. In: *Proc. IJCAI-77,* pp. 920–927 (Cambridge, MA, Aug. 22–25, 1977). MIT, Cambridge, MA, 1977.

[de Hoog et al., 1992] R. de Hoog, R. Martil, B. Wielinga, R. Taylor, C. Bright, and W. van de Velde. *The Common KADS Model Set.* Research Report ESPRIT-Project P5248 KADS-II/WPI- II/RR/UvA/018/4.0, University of Amsterdam, Amsterdam, 1992.

[Dean & Boddy, 1988] T. L. Dean and M. Boddy. An Analysis of Time-Dependent Planning. In: *Proc. AAAI-88,* pp. 49–54 (St. Paul, MN, Aug., 1988). Morgan Kaufmann, Palo Alto, CA, 1988.

[Dean & McDermott, 1987] T. L. Dean and D. V. McDermott. Temporal Data Base Management. *Journal of Artificial Intelligence 32,* pp. 1–55, 1987.

[De Raedt & Bruynooghe, 1992] L. De Raedt and M. Bruynooghe. Interactive Concept-Learning and Constructive Induction by Analogy. *Machine Learning 8*(2), pp. 107–150, 1992.

[De Raedt & Lavrac, 1993] L. De Raedt and N. Lavrac. The Many Faces of Inductive Logic Programming. In: J. Komorowski, editor, *Proceedings of the 7th International Symposium on Methodologies for Intelligent Systems,* (Trondheim, Norway, June 15–18, 1993). Lecture Notes in Artificial Intelligence. Springer-Verlag, Berlin, 1993.

[di Primio, 1991] F. di Primio. *Verarbeitung ungenauer Anweisungen anhand des Beispiels Raumeinrichtung.* TASSO-Report 26, GMD, Sankt Augustin, August 1991.

[di Primio, 1993] F. di Primio. *Hybride Wissensverarbeitung.* DUV/Vieweg, Wiesbaden, 1993.

[Dierkes et al., 1991] M. Dierkes, U. Hoffmann, and L. Marz. *Leitbilder und Technik. Zur Entstehung und Steuerung technischer Innovation.* Edition Sigma, Berlin, 1991.

[Dörner, 1989] D. Dörner. *Die Logik des Mißlingens.* Rowohlt, Reinbek, 1989.

[Dourish & Bellotti, 1992] P. Dourish and V. Bellotti. Awareness and Coordination in Shared Workspaces. In: *Proceedings of the Conference on Computer-Supported Cooperative Work,* pp. 25–38 (November 1992). ACM Press, New York, 1992.

[Dzeroski & Bratko, 1992] S. Dzeroski and I. Bratko. Handling Noise in Inductive Logic Programming. *Proceedings of Second International Workshop on Inductive Logic Programming* (Technical Report TM-1182). ICOT, Tokyo, Japan, 1992.

[Edmonds, 1987] E. Edmonds. Adaptation, Response and Knowledge. *Knowledge Based Systems 1*(1), pp. 3–10, 1987.

[Eisenberg & Fischer, 1993] M. Eisenberg and G. Fischer. Learning on Demand—Why Is It Necessary and Why Does It Make a Difference? In: *Proceedings of the Fifteenth Annual Conference of the Cognitive Science Society,* pp. 180–181 (Boulder, CO, June 18–21, 1993). Lawrence Erlbaum Associates, Hillsdale, NJ, 1993.

[Ellis, 1974] W. D. Ellis. *A Source Book of Gestalt Psychology.* 5th impression, first published 1938, Lowe & Brydone, Thetford, UK, 1974.

[Emde, 1989] W. Emde. An Inference Engine for Representing Multiple Theories. In: K. Morik, editor, *Knowledge Representation and Organization in Machine Learning,* pp. 148–176. Springer-Verlag, New York, 1989.

[Emde, 1991] W. Emde. *Modellbildung, Wissensrevision und Wissensrepräsentation im Maschinellen Lernen.* Dissertation. Informatik-Fachberichte 281, Springer-Verlag, New York, 1991.

[Emde, 1994] W. Emde. Inductive Learning of Characteristic Concept Descriptions from Small Sets of Classified Examples. In: F. Bergadano and L. De Raedt, editors, *Machine Learning: ECML-94, European Conference on Machine Learning*, pp. 103–121 (Catania, Italy, April 1994). Springer-Verlag, Berlin, 1994. Also as Arbeitspapiere der GMD 821.

[Emde et al., 1993] W. Emde, J. Kietz, K. Morik, E. Sommer, and S. Wrobel. *MOBAL 2.0 User Guide.* Arbeitspapiere der GMD 777, GMD, Sankt Augustin, 1993.

[EN 29241] EN 29241. *Ergonomische Anforderungen für Bürotätigkeiten mit Bildschirmgeräten.*

[Engelbart, 1990] D. Engelbart. Knowledge-Domain Interoperability and an Open Hyperdocument System. In: *Proc. CSCW'90*, pp. 143–156 (Los Angeles, Oct. 7–10, 1990). ACM Press, New York, 1990.

[Englert, 1995] R. Englert. Knowledge-Domain Interoperability and an Open Hyperdocument System. In: *Proc. 17. Fachtagung für Künstliche Intelligenz (KI-95)*, pp. 77–88 (Bielefeld, Sept. 11–13, 1995). Springer-Verlag, Berlin, 1995.

[Etherington et al., 1989] D. Etherington, A. Borgida, R. J. Brachman, and H. Kautz. Vivid Knowledge and Tractable Reasoning: Preliminary Report. In: *Proc. IJCAI-89*, pp. 1146–1152 (Detroit, MI, Aug. 20–25, 1989). Morgan Kaufmann, San Mateo, CA, 1989.

[Fahlen et al., 1993] L. E. Fahlen, C. G. Brown, O. Stahl, and C. Carlsson. A Space-Based Model for User Interaction in Shared Synthetic Environments. In: *Proceedings of the CHI Conference on Human Factors in Computing Systems*, pp. 43–48 (Amsterdam, April 24–29, 1993). ACM Press, New York, 1993.

[Feiner, 1988] S. Feiner. An Architecture for Knowledge-Based Graphical Interfaces. In: *Proceedings of the ACM SIGCHI Workshop on architectures of intelligent interfaces*, pp. 129–140. Monterey, CA, 1988.

[Feldman, 1987] J. A. Feldman. A Functional Model of Vision and Space. In: M. A. Arbib and A. R. Hanson, editors, *Vision, Brain, and Cooperative Computation*, pp. 531–562. MIT Press, Cambridge, MA, 1987.

[Fellner & Stögerer, 1988] W. D. Fellner and J. K. Stögerer. PIC—eine objektorientierte grafische Abfragesprache. In: *Proc. Austrographics '88*, pp. 85–101 (Vienna, Austria, Sept. 28–30, 1988). Springer-Verlag, Vienna, Austria, 1988.

[Fikes & Nilsson, 1971] R. E. Fikes and N. J. Nilsson. STRIPS: A New Approach to Theorem Proving in Problem Solving. *Journal of Artificial Intelligence 2*, pp. 189–208, 1971.

[Fikes et al., 1972] R. E. Fikes, P. E. Hart, and N. J. Nilsson. Learning and Executing Generalized Robot Plans. *Journal of Artificial Intelligence 3*, pp. 251–288, 1972.

[Fischer & Lemke, 1988] G. Fischer and A. Lemke. Construction Kits and Design Environments: Steps Toward Human Problem-Domain Communication. *HUMAN–COMPUTER INTERACTION 1987–1988*(3), pp. 179–222, 1988.

[Fischer et al., 1991] G. Fischer, A. Lemke, T. Mastaglio, and A. Morch. The Role of Critiquing in Cooperative Problem Solving. *ACM Transactions on Information Systems 9*(3), pp. 123–151, 1991.

[Flavell, 1984] J. H. Flavell. Annahmen zum Begriff Metakognition sowie zur Entwicklung von Metakognition. In: F. Weinert and R. Kluwe, editors, *Metakognition, Motivation und Lernen*, pp. 23–30. Kohlhammer, Stuttgart, 1984.

[Fowler et al., 1987] C. J. H. Fowler, L. A. Macaulay, and S. Siripoksup. An Evaluation of the Effectiveness of the Adaptive Interface Module (AIM) in Matching Dialogues to Users. In: D. Diaper and R. Winder, editors, *People and Computers III*, pp. 346–359. Cambridge University Press, Cambridge, 1987.

[Freeman-Benson et al., 1990] B. N. Freeman-Benson, J. Maloney, and A. Borning. An Incremental Constraint Solver. *Communications of the ACM 33*, pp. 54–63, 1990.

[Gebhardt, 1991] F. Gebhardt. Choosing Among Competing Generalizations. *Knowledge Acquisition 3*, pp. 361–380, 1991.

[Gebhardt, 1994] F. Gebhardt. Interessantheit als Kriterium für die Bewertung von Ergebnissen. *Informatik Forschung und Entwicklung 9*, pp. 9–21, 1994.

[Geffner & Pearl, 1992] H. Geffner and J. Pearl. Conditional Entailment: Bridging Two Approaches to Default Reasoning. *Artificial Intelligence 53*(2–3), pp. 209–244, 1992.

[Gentner, 1982] D. Gentner. Are Scientific Analogies Metaphors? In: D. S. Miall, editor, *Metaphor*, pp. 106–132. Harvester, Brighton, UK, 1982.

[Gentner, 1983] D. Gentner. Structure-Mapping: A Theoretical Framework for Analogy. *Cognitive Science 7*, pp. 155–170, 1983.

[Gillies & Khan, 1992] D. F. Gillies and G. N. Khan. Perceptual Grouping and the Hough Transform. In: *Proceedings of the SPIE—The International Society for Optical Engineering*, vol. 1607, pp. 188–196, 1992.

[Ginsberg & Smith, 1988] M. L. Ginsberg and D. E. Smith. Reasoning About Action I: A Possible Worlds Approach. *Journal of Artificial Intelligence 35*, pp. 165–195, 1988.

[Gordon, 1993] T. F. Gordon. The Pleadings Game: Formalizing Procedural Justice. In: *Proceedings of the Fourth International Conference on Artificial Intelligence and Law*, pp. 10–19. ACM Press, New York, 1993.

[Gordon, 1995] T. F. Gordon. *The Pleadings Game: An Artificial Intelligence Model of Procedural Justice*. Kluwer Academic Publishers, Dordrecht, 1995. Also as: Ph.D. diss., Fachbereich Informatik, Technische Hochschule Darmstadt, 1993.

[Gordon et al., 1993] T. F. Gordon, J. Hertzberg, and A. Horz. *A Planner for Beautifying Business Graphics*. TASSO Report 45, GMD, Sankt Augustin, 1993.

[Gordon et al., 1994] T. F. Gordon, J. Hertzberg, and A. Horz. *The qwertz Toolbox Reference Manual*. Arbeitspapiere der GMD 835, GMD, Sankt Augustin, 1994.

[Grice, 1975] H. P. Grice. Logic and Conversation. In: P. Cole and J. L. Morgan, editors, *Syntax and Semantics: Vol. 3, "Speech Acts"*, pp. 43–58. Academic Press, New York, 1975.

[Grob, 1985] R. Grob. *Flexibilität in der Fertigung*. Springer-Verlag, Berlin, 1985.

[Grønbæk et al., 1992] K. Grønbæk, M. Kyng, and P. Mogensen. CSCW Challenges in Large-Scale Technical Projects: A Case Study. In: J. Turner and R. Kraut, editors, *CSCW'92*, Proceedings of the Conference on Computer-Supported Cooperative Work, pp. 338–345 (Toronto, Canada, Oct. 31–Nov. 4, 1992). ACM Press, New York, 1992.

[Grudin, 1990] J. Grudin. Groupware and Cooperative Work: Problems and Prospects. In: B. Laurel, editor, *The Art of Human Computer Interface Design*, pp. 171–175. Addison-Wesley, Reading, MA, 1990.

[Grunst, 1988] G. Grunst. *Videoanalysen einiger Nutzungsprobleme von Word 3 für den IBM-PC und den Macintosh. Eine Pilotuntersuchung*. Arbeitspapiere der GMD 307, GMD, Sankt Augustin, 1988.

[Grunst, 1993] G. Grunst. Adaptive Hypermedia for Support Systems. In: M. Schneider-Hufschmidt, T. Kühme, and U. Malinowski, editors, *Adaptive User Interfaces*, pp. 269–283. Elsevier Science Publishers, New York, 1993.

[Guiliano et al., 1961] V. E. Guiliano, P. E. Jones, G. E. Kimball, R. F. Meyer, and B. A. Stein. Automatic Pattern Recognition by a Gestalt Method. *Information Control 4*, pp. 332–345, 1961.

[Güsgen & Fidelak, 1990] H. W. Güsgen and M. Fidelak. Towards Reasoning about Space and Time. In: *Räumliche Alltags-Umgebungen des Menschen (RAUM)*, pp. 53–68, Universität Koblenz, Koblenz, 1990.

[Haake & Wilson, 1992] J. M. Haake and B. Wilson. Supporting Collaborative Writing of Hyperdocuments in SEPIA. In: *Proceedings of the Conference on Computer-Supported Cooperative Work*, pp. 138–146 (Nov. 1992). ACM Press, New York, 1992.

[Habel, 1983] C. Habel. Logische Systeme und Repräsentationsprobleme. In: B. Neumann, editor, *Proc. GWAI-83*, pp. 118–142 (Dassel, Germany, Sept. 19–23, 1983). Springer-Verlag, Berlin, 1983.

[Habel, 1990] C. Habel. Propositional and Depictorial Representations of Spatial Knowledge: The Case of *path*-Concepts. In: R. Studer, editor, *Natural Language and Logic. Proceedings of the International Scientific Symposium*, pp. 94–117 (Hamburg, May 9–11, 1989). Springer-Verlag, Berlin, 1990.

[Haller, 1974] F. Haller. *MIDI—ein offenesSystem für mehrgeschossige Bauten mit integrierter Medieninstallation*. USM Bausysteme Haller, Münsingen, 1974.

[Haller, 1985] F. Haller. *ARMILLA—Ein Installationsmodell*. IFIB, Universität Karlsruhe, Karlsruhe, 1985.

[Hart, 1961] H. L. A. Hart. *The Concept of Law*. Oxford University Press, Oxford, UK, 1961.

[Hartson & Boehm-Davis, 1993] H. Hartson and D. Boehm-Davis. User Interface Development Processes and Methodologies. *Behaviour & Information Technology 12*(2), pp. 98–114, 1993.

[Hayes et al., 1981] P. Hayes, E. Ball, and R. Reddy. Breaking the Man–Machine Communication Barrier. *IEEE Computer 14*, pp. 3–30, 1981.

[Helft, 1989] N. Helft. Induction as Nonmonotonic Inference. In: *Proceedings of the 1st International Conference on Knowledge Representation and Reasoning*. 1989.

[Helm, 1992] G. Helm. *Metaphern in der Informatik. Begriffe, Theorien, Prozesse*. Arbeitspapiere der GMD 652, GMD, Sankt Augustin, 1992.

[Helson, 1933] H. H. Helson. The Fundamental Propositions of Gestaltpsychologie. *Psychological Review 70*, pp. 13–32, 1933.

[Henderson & Stuart, 1986] D. A. Henderson and K. C. Stuart. Rooms: The Use of Multiple Virtual Workspaces to Reduce Space Contention in a Window-Based Graphical User Interface. *ACM Transactions on Graphics 5*(3), pp. 211–243, July 1986.

[Henne, 1990] P. Henne. *Ein experimentelles Assoziativspeicher-Modell*. TASSO Report 12, GMD, Sankt Augustin, November 1990.

[Henne & Schmitgen, 1991] P. Henne and G. Schmitgen. *Grafische Suche: Benutzerwünsche, Einsatzbereiche, Repräsentationsformen, Methoden*. TASSO Report 34, GMD, Sankt Augustin, 1991.

[Henne & Schmitgen, 1993] P. Henne and G. Schmitgen. *Grafisches Suchsystem SkaGra: Konzepte und Realisierungen*. TASSO Report 49, GMD, Sankt Augustin, 1993.

[Hennessy et al., 1992] P. Hennessy, T. Kreifelts, and U. Ehrlich. Distributed Work Management: Activity Coordination within the EuroCoOp Project. *Computer Communications 15*(8), pp. 477–488, 1992.

[Hernandez, 1990] D. Hernandez. Using Comparative Relations to Represent Spatial Knowledge. In: *Räumliche Alltags-Umgebungen des Menschen (RAUM)*, pp. 69–80. Universität Koblenz, Koblenz, 1990.

[Hertzberg, 1989] J. Hertzberg. *Planen. Einführung in die Planerstellungsmethoden der Künstlichen Intelligenz*. BI Wissenschaftsverlag, Mannheim, 1989.

[Hertzberg, 1991] J. Hertzberg. *Revising Planning Goals.* TASSO Report 3, GMD, Sankt Augustin, March 1990.

[Hertzberg, 1993] J. Hertzberg. KI-Handlungsplanung—Woran wir arbeiten, und woran wir arbeiten sollten. In: O. Herzog, T. Christaller, and D. Schütt, editors, *Grundlagen und Anwendungen der Künstlichen Intelligenz. 17. Fachtagung für Künstliche Intelligenz (KI-93)*, pp. 3–27 (Berlin, Sept. 13–16, 1993). Springer-Verlag, Berlin, 1993.

[Hertzberg, 1994] J. Hertzberg. Theoretical Planning and its Contributions to Practical and Applied Planning (extended abstract). In: A. G. Cohn, editor, *11th European Conference on Artificial Intelligence ECAI-94*, pp. 811–812 (Amsterdam, Aug. 8–12, 1994). Wiley & Sons, Amsterdam, 1994.

[Hertzberg & Horz, 1989] J. Hertzberg and A. Horz. Towards a Theory of Conflict Detection and Resolution in Nonlinear Plans. In: *Proc. IJCAI-89*, pp. 937–942 (Detroit, MI, Aug. 20–25, 1989). Morgan Kaufmann, San Mateo, CA, 1989.

[Hertzberg & Thiébaux, 1994] J. Hertzberg and S. Thiébaux. Turning an Action Formalism into a Planner—A Case Study. *Journal of Logic and Computation 4(5)*, pp. 617–654, 1994.

[Hill & Miller, 1988] W. C. Hill and J. R. Miller. Justified Advice: A Seminaturalistic Study of Advisory Strategies. In: E. Soloway, D. Frye, and S. B. Sheppard, editors, *Proceedings of CHI'88: Human Factors in Computing Systems*, pp. 185–190 (Washington, DC, May 15–19, 1988). ACM Press, New York, 1988.

[Horacek, 1991a] H. Horacek. Exploiting Conversational Implicature for Generating Concise Explanations. In: *Proceedings of the Fifth Conference of the European Chapter of the Association for Computational Linguistics (EACL-91)*, pp. 191–193 (Morristown, NJ, April 9–11, 1991). ACL, Berlin, 1991.

[Horacek, 1991b] H. Horacek. Towards Finding the Reasons Behind—Generating the Content of Explanation. In: T. Christaller, editor, *GWAI-91, 15th German Workshop on Artificial Intelligence*, pp. 96–105 (Bonn, Dec. 16–20, 1991). Springer-Verlag, Berlin, 1991.

[Horacek, 1992a] H. Horacek. *Natural Language Explanation for Constraint-Based Expert Systems.* DIAMOD Report 21, Universität Bielefeld, Fakultät für Linguistik und Literaturwissenschaft, December 1992.

[Horacek, 1992b] H. Horacek. An Integrated View of Text Planning. In: R. Dale, D. Roesner, and O. Stock, editors, *Aspects of Automated Natural Language Generation. Proceedings of the 6th International Workshop on Natural Language Generation*, pp. 29–44 (Trento, Italy, Apr. 1992). Springer-Verlag, New York, 1992.

[Horz, 1994] A. Horz. Relating Classical and Temporal Planning (Preliminary Report). In: C. Bäckström and E. Sandewall, editors, *Current Trends in AI Planning. EWSP'93—2nd European Workshop on Planning*, pp. 145–157 (Vadstena, Sweden, Dec. 9–11, 1993). IOS Press, Amsterdam, 1994.

[Hoschka, 1991] P. Hoschka. Assisting-Computer—A New Generation of Support Systems. In: W. Brauer and D. Hernandez, editors, Verteilte Künstliche Intelligenz und Kooperatives Arbeiten. *Proc. 4. Int. GI-Kongress Wissensbasierte Systeme*, pp. 219–230 (Munich, Oct. 23–24, 1991). Springer-Verlag, New York, 1991.

[Hoschka & Klösgen, 1991] P. Hoschka and W. Klösgen. A Support System for Interpreting Statistical Data. In: G. Piatetsky-Shapiro and W. Frawley, editors, *Knowledge Discovery in Databases*, pp. 325–346. MIT Press, Cambridge, MA, 1991.

[Hoschka et al., 1994] P. Hoschka, T. Kreifelts, and W. Prinz. Gruppenkoordination und Vorgangsbearbeitung. In: S. Kirn and K. Klöckner, editors, *Betrieblicher Einsatz von CSCW-Systemen*, pp. 91–112. GMD, Sankt Augustin, 1994.

[Hovestadt, 1993] L. Hovestadt. A4—An Integrated Building Model Based on Visualisation. In: M. R. Beheshti and K. Zreik, editors, *Advanced Technologies—Architecture—Planning—Civil Engineering, Fourth EuropIA International Conference on the Application of Artificial Intelligence, Robotics and Image*, pp. 265–296. Delft, June 1993.

[Hovestadt, 1994] L. Hovestadt. *A4—Digitales Bauen—Ein Modell für die weitgehende Computerunterstützung von Entwurf, Konstruktion und Betrieb von Gebäuden.* Forschrittberichte VDI, Reihe 20: Rechnerunterstützte Verfahren Bd. 120. VDI, Düsseldorf, 1994. Also as: Dissertation, Universität Karlsruhe (TU), Institut für Industrielle Bauproduktion, Karlsruhe.

[Hudlicka & Lesser, 1984] E. Hudlicka and V. R. Lesser. Meta-Level Control through Fault Detection and Diagnosis. In: *Proceedings of the National Conference on Artificial Intelligence (AAAI-84)*, pp. 153–161 (Austin, TX, Aug. 6–10, 1984). Morgan Kaufmann, Los Altos, CA, 1984.

[IBM, 1992] IBM. *Object-Oriented Interface Design, IBM Common User Access Guidelines.* Que Corporation, Carmel, 1992.

[Indurkhya, 1992] B. Indurkhya. *Metaphor and Cognition.* Kluwer, Boston, 1992.

[Ingendahl, 1973] W. Ingendahl. *Der metaphorische Prozeß.* Schwann, Düsseldorf, 1973.

[ISO 9241] ISO 9241. *Ergonomic Requirements for Office Work with Visual Display Terminals.* ISO International Organization for Standardization, Geneve, Switzerland.

[Jansweijer, 1988] W. Jansweijer. *PDP.* Ph.D. thesis, University of Amsterdam, Amsterdam, 1988.

[Johnson et al., 1980] J. R. Johnson, R. R. Rice, and R. A. Roemmich. Pictures That Lie: The Abuse of Graphs in Annual Reports. *Management and Accounting 62*(4), pp. 50–60, 1980.

[Johnson-Laird, 1983] P. N. Johnson-Laird. *Mental Models—Towards a Cognitive Science of Language, Inference, and Consciousness.* Harvard University Press, Cambridge, MA, 1983.

[Johnson-Laird, 1988] P. N. Johnson-Laird. *The Computer and the Mind.* Harvard University Press, Cambridge, MA, 1988.

[Jung, 1976] C. G. Jung. *Die Archetypen und das kollektive Unbewußte.* Walter-Verlag, Freiburg im Breisgau, 1976.

[Junker, 1991] U. Junker. *The EXCEPT II Default Reasoning System.* TASSO Report 23, GMD, Sankt Augustin, 1990.

[Kahn et al., 1990] P. Kahn, A. Winkler, and C. Y. Chong. Perceptual Grouping as Energy Minimization. In: *1990 IEEE International Conference on Systems, Man and Cybernetics*, pp. 542–546 (Los Angeles, Nov. 4–7, 1990). IEEE, New York, 1990.

[Kansy, 1990] K. Kansy. *Leitbeispiel Graphikdesigner.* TASSO Report 14, GMD, Sankt Augustin, 1990.

[Kansy, 1991] K. Kansy. Der Grafikdesigner—wissensbasiertes Erzeugen und Verschönern von Grafiken. *Office Management 39*(10), pp. 25–28, 1991.

[Karbach & Voß, 1992a] W. Karbach and A. Voß. Glueing Together Small Solutions—An Office Planning Model in MODEL-K. In: M. Linster, editor, *Sisyphus-91: Models of Problem Solving*, paper 7. Arbeitspapiere der GMD 663, GMD, Sankt Augustin, 1992.

[Karbach & Voß, 1992b] W. Karbach and A. Voß. Reflecting about Expert Systems in MODEL-K. In: J. C. Rault, editor, *Proceedings of AVIGNON-92*, pp. 141–152. 1992.

[Karbach et al., 1990] W. Karbach, M. Linster, and A. Voß. Models, Methods, Roles and Tasks: Many Labels—One Idea? *Knowledge Acquisition Journal 2*, pp. 279–299, 1990.

[Kato et al., 1992] T. Kato, T. Kurita, N. Otsu, and K. Hirata. A Sketch Retrieval Method for Full Color Image Database. In: *Proceedings 11th IAPR International Conference on Pattern Recognition*, pp. 530–533. IEEE Computer Society Press, Los Alamitos, CA, 1992.

[Keene, 1989] S. E. Keene. *Object-Oriented Programming in COMMON LISP—A Programmer's Guide to CLOS*. Addison-Wesley, Reading, MA, 1989.

[Kemke, 1992] C. Kemke. Metaphern und analogisches Schließen. Ein Überblick. *KI—Künstliche Intelligenz 2/1992*, pp. 28–36, 1992.

[Kendall, 1992] J. E. Kendall. Using Metaphors to Enhance Intelligence in Information Systems. Rationale for an Alternative to Rule-Based Intelligence. In: Vogt, editor, *Personal Computers and Intelligent Systems. IFIP Transactions A-14*, pp. 213–219. North Holland, Amsterdam, 1992.

[Kendall & Kendall, 1992] J. E. Kendall and K. E. Kendall. *Metaphors and Methodologies. Living Beyond the Systems Machine*. Manuscript, Camden School of Business, Rutgers University, Camden, NJ, Aug. 1992. Also as: *MIS-Quaterly 17*(2), pp. 149-171, June 1993.

[Kieras & Polson, 1985] D. E. Kieras and P. G. Polson. An Approach to the Formal Analysis of User Complexity. *International Journal of Man–Machine Studies 22*, pp. 365–394, 1985.

[Kietz, 1988] J.-U. Kietz. Incremental and Reversible Acquisition of Taxonomies. In: J. Boose, editor, *Proc. EKAW-88*, pp. 24.1–24.11 (Bonn, June 19–23, 1988). GMD-Studien 143, GMD, Sankt Augustin, 1988. Also as: KIT-Report 66, Technical University of Berlin, Berlin, 1988.

[Kietz & Wrobel, 1992] J.-U. Kietz and S. Wrobel. Controlling the Complexity of Learning in Logic through Syntactic and Task-Oriented Models. In: S. Muggleton, editor, *Inductive Logic Programming*, chapter 16. Academic Press, London, 1992.

[Kille, 1989] S. Kille. The Quipu Directory Service. In: *Proceedings of Message Handling Systems and Distributed Applications, IFIP 6.5 International Working Conference*, pp. 173–186 (Costa Mesa, CA, Oct. 10–12, 1988). North Holland, New York, 1989.

[Klingspor, 1991] V. Klingspor. *MLT deliverable 4.3.2/G: Mobal's Predicate Structuring Tool*. Arbeitspapiere der GMD 592, GMD, Sankt Augustin, September 1991.

[Klingspor, 1994] V. Klingspor. GRDT: Enhancing Model-Based Learning for its Application in Robot Navigation. In: S. Wrobel, editor, *Proceedings of the 4th International Workshop on Inductive Logic Programming (ILP-94)*, pp. 107–122 (Bad Honnef, Germany, Sept. 12–14, 1994). GMD-Studien 237, GMD, Sankt Augustin, 1994.

[Klinker et al., 1990] G. Klinker, C. Bhola, G. Dallemange, D. Marques, and J. McDermott. Usable and Reusable Programming Constructs. In: J. H. Boose and B. R. Gaines, editors, *Proc. KAW-90*, pp. 14-1–14-20. University of Calgary, Calgary, Canada, 1990.

[Klösgen, 1992] W. Klösgen. Problems for Knowledge Discovery in Databases and their Treatment in the Statistics Interpreter EXPLORA. *International Journal for Intelligent Systems 7*(7), pp. 649–673, 1992.

[Klösgen, 1993] W. Klösgen. *Explora: A Support System for Discovery in Databases, Version 1.1, User Manual*. GMD, Sankt Augustin, 1993.

[Klösgen, 1995a] W. Klösgen. Efficient Discovery of Interesting Statements in Databases. *Journal of Intelligent Information Systems 4*, pp. 1–17, 1995.

[Klösgen, 1995b] W. Klösgen. Explora: A Multipattern and Multistrategy Discovery Assistant. In: U. Fayyad, G. Piatetsky-Shapiro, P. Smyth, and R. Uthurusamy, editors, *Advances in Knowledge Discovery and Data Mining*. MIT Press, Cambridge, MA, 1995, Forthcoming.

[Klösgen & Quinke, 1985] W. Klösgen and H. Quinke. Sozioökonomische Simulation und Planung: Entwicklungsstand und Computerunterstützung. *Informatik-Spektrum 8*, pp. 328–336, 1985.

[Klösgen & Schwarz, 1983] W. Klösgen and W. Schwarz. *Modellbank-System MBS (User Manual)*. Arbeitspapiere der GMD 32, GMD, Sankt Augustin, 1983.

[Klösgen & Zytkow, 1995] W. Klösgen and J. Zytkow. Knowledge Discovery in Databases Terminology. In: U. Fayyad, G. Piatetsky-Shapiro, P. Smyth, and R. Uthurusamy, editors, *Advances in Knowledge Discovery and Data Mining*. MIT Press, Cambridge, MA, 1995, Forthcoming.

[Kluwe & Schiebler, 1984] R. H. Kluwe and K. Schiebler. Entwicklung exekutiver Prozesse und kognitiver Leistungen. In: F. Weinert and R. Kluwe, editors, *Metakognition, Motivation und Lernen*, pp. 31–59. Kohlhammer, Stuttgart, 1984.

[Köhler, 1929] W. Köhler. *Gestalt Psychology*. Liveright, New York, 1929.

[Kolb, 1992] R. Kolb. *Die graphische Umsetzung statistischer Aussagen*. TASSO Report 39, GMD, Sankt Augustin, 1992.

[Kolb, 1993] R. Kolb. *Applying Non-Monotonic Reasoning to Graphics Design*. TASSO Report 52, GMD, Sankt Augustin, 1993.

[Kolodner, 1993] J. L. Kolodner. *Case-Based Reasoning*. Morgan Kaufmann, San Mateo, CA, 1993.

[Krasner & Pope, 1988] G. E. Krasner and S. T. Pope. A Cookbook for Using the Model-View-Controller User Interface Paradigm in Smalltalk-80. *Journal of Object-Oriented Programming 1*(3), pp. 26–49, 1988.

[Kreifelts & Prinz, 1993] T. Kreifelts and W. Prinz. ASCW—An Assistant for Cooperative Work. In: *Proceedings of the Conference on Organizational Computing Systems*, pp. 269–278 (Milpitas, CA, Nov. 1–4). ACM Press, New York, 1993.

[Kreifelts et al., 1984] T. Kreifelts, U. Licht, P. Seuffert, and G. Woetzel. DOMINO: A System for the Specification and Automation of Cooperative Office Processes. In: B. Myrhaug and D. R. Wilson, editors, *Advances in Microprocessing and Microprogramming, Proc. EUROMICRO '84*, pp. 33–41 (Copenhagen, Denmark, Aug. 27–30, 1984). North Holland, Amsterdam, 1984.

[Kreifelts et al., 1991] T. Kreifelts, E. Hinrichs, K. H. Klein, P. Seuffert, and G. Woetzel. Experiences with the DOMINO Office Procedure System. In: L. Bannon, M. Robinson, and K. Schmidt, editors, *ECSCW'91, Proceedings of the Second European Conference on Computer-Supported Cooperative Work*, pp. 117–130 (Amsterdam, Sept. 24–27, 1991). Kluwer, Dordrecht, 1991.

[Kreifelts et al., 1993] T. Kreifelts, E. Hinrichs, and G. Woetzel. Sharing To-Do Lists with a Distributed Task Manager. In: G. DeMichelis, editor, *ECSCW'93, Proceedings of the Third European Conference on Computer-Supported Cooperative Work*, pp. 31–46 (Milan, Italy, Sept. 13–17, 1993). Kluwer, Dordrecht, 1993.

[Kuhn, 1962] T. S. Kuhn. *The Structure of Scientific Revolution*. University of Chicago Press, Chicago, second edition 1970.

[Kyng & Greenbaum, 1991] M. Kyng and J. Greenbaum, editors. *Design at Work*. Lawrence Erlbaum Associates, London, UK, 1991.

[Lakoff, 1987] G. Lakoff. *Women, Fire, and Dangerous Things*. University of Chicago Press, Chicago, 1987.

[Lakoff & Johnson, 1980] G. Lakoff and M. Johnson. *Metaphors We Live By.* University of Chicago Press, Chicago, 1980.

[Levergood et al., 1993] T. M. Levergood, A. C. Payne, J. Gettys, G. W. Treese, and L. C. Stewart. AudioFile: A Network-Transparent System for Distributed Audio Applications. In: *Proceedings of the USENIX Summer Conference,* pp. 219–236 (Cincinnati, OH, June 21–25, 1993). USENIX, Cincinatti, OH, 1993.

[Li & Mantei, 1992] J. Li and M. Mantei. Working Together, Virtually. In: *Proceedings of Graphics Interface,* pp. 115–122 (Vancouver, BC, Canada, May 11–15, 1992). 1992.

[Lifschitz, 1987] V. Lifschitz. On the Semantics of STRIPS. In: M. P. Georgeff and A. L. Lansky, editors, *Proceedings of the 1986 Workshop Reasoning about Actions and Plans,* pp. 1–9 (Timberline, OR, June 30–July 2, 1986). Morgan Kaufmann, Los Altos, CA, 1987.

[Lindner, 1994] G. Lindner. *Logikbasiertes Lernen in relationalen Datenbanken.* LS-8 Report, Universität Dortmund, FB Informatik, Dortmund, 1994.

[Löwgren & Nordquist, 1992] J. Löwgren and T. Nordquist. Knowledge-Based Evaluation as Design Support for Graphical User Interfaces. In: *Proc. CHI'92,* pp. 181–188. Addison-Wesley, Reading, MA, 1992.

[Luhmann, 1984] N. Luhmann. Soziale Systeme. Grundriß einer allgemeinen Theorie. Suhrkamp, Frankfurt am Main, 1984. An English introduction to Luhmann's work is: N. Luhmann, *Essays in Self-Realization,* Columbia University Press, New York, 1990.

[Maaß, 1992] W. Maaß. CLAY—Ein System zur constraint-basierten Plazierung multimodaler Objekte in Dokumenten. In: K. Kansy and P. Wißkirchen, editors, *Innovative Programmiermethoden für Graphische Systeme,* pp. 45–56. Springer-Verlag, Berlin, 1992.

[Mackinlay, 1986] J. D. Mackinlay. Automating the Design of Graphical Presentations of Relational Information. *ACM TOG 5,* pp. 110–141, 1986.

[Maes, 1987] P. Maes. *Computational Reflection* (Technical Report 87–2). Free University of Brussels, AI Lab, Brussels, Belgium, 1987.

[Maes, 1988] P. Maes. Issues in Computational Reflection. In: P. Maes and D. Nardi, editors, *Meta-Level Architectures and Reflection,* pp. 21–35. North Holland, Amsterdam, 1988.

[Mambrey, 1994] P. Mambrey. Die technische Metapher als Kommunikationsmedium und Konstruktionshilfe. In: *Technik und Gesellschaft. Jahrbuch 7: Konstruktion und Evolution von Technik,* pp. 127–148. Campus Verlag, New York, 1994.

[Mambrey, 1995] P. Mambrey. Metaphors as Requirement Analysis Tools. The Market Metaphor in CSCW System Design. In: D. Shapiro, editor, *Design of Computer Supported Cooperative Work and Groupware Systems.* North Holland, Amsterdam and New York, 1995, Forthcoming.

[Mambrey et al., 1994] P. Mambrey, M. Paetau, and A. Tepper. Controlling Visions And Metaphors. In: K. A. Duncan and K. Krüger, editors, *Linkage and Development Countries. Proceedings of the IFIP 13th World Computer Congress,* vol. III, pp. 223–228 (Hamburg, Aug. 28–Sept. 2, 1994). North Holland, Amsterdam, 1994.

[Mambrey & Tepper, 1992] P. Mambrey and A. Tepper. *Metaphern und Leitbilder als Instrument. Beispiele und Methoden.* Arbeitspapiere der GMD 651, GMD, Sankt Augustin 1992.

[Markus & Connolly, 1990] K. Markus and T. Connolly. Why CSCW Applications Fail: Problems in the Adoption of Interdependent Work Tools. In: *Proc. CSCW'90,* pp. 371–380 (Los Angeles, Oct. 7–10, 1990). ACM Press, New York, 1990.

[Marshall, 1989] C. C. Marshall. Representing the Structure of Legal Argument. In: *Proceedings of the Second International Conference on Artificial Intelligence and Law*, pp. 121–127 (Vancouver, BC, Canada, June 13–16, 1989). ACM Press, New York, 1989.

[Martin, 1990] J. H. Martin. *A Computational Model of Metaphor Interpretation.* Academic Press, San Diego, 1990.

[Mayhew, 1992] D. Mayhew. *Principles and Guidelines in Software User Interface Design.* Prentice Hall, Englewood Cliffs, NJ, 1992.

[McCafferty, 1990] J. D. McCafferty. *Human and Machine Vision: Computing Perceptual Organisation.* Ellis Horwood, Chichester, UK, 1990.

[McDermott, 1992] D. McDermott. Robot Planning. *AI Magazine 13*(2), pp. 55–79, 1992.

[McKeown, 1985] K. R. McKeown. *Textgeneration.* Cambridge University Press, Cambridge, 1985.

[Meier, 1991] J. Meier. *Erklärung in KADS und EES.* DIAMOD Report 8, Universität Bielefeld, Fakultät für Linguistik und Literaturwissenschaft, Bielefeld, September 1991.

[Meier, 1992] J. Meier. *A Semantic View of Explanation.* DIAMOD Report 15, Universität Bielefeld, Fakultät für Linguistik und Literaturwissenschaft, Bielefeld, Sept. 1992.

[Michalski, 1993] R. S. Michalski. Inferential Theory of Learning as a Conceptual Basis for Multistrategy Learning. *Machine Learning 11*(2/3), pp. 111–152, 1993.

[Microsoft, 1992] Microsoft. *The Windows Interface, An Application Design Guide.* Microsoft Press, Redmond, WA, 1992.

[Minsky, 1980] M. Minsky. A Framework for Representing Knowledge. In: D. Metzing, editor, *Frame Conception and Text Understanding*, pp. 1–25. W. de Gruyter, Berlin, 1980.

[Mitchell & Shneiderman, 1989] J. Mitchell and B. Shneiderman. Dynamic versus Static Menus: An Exploratory Comparison. *SIGCHI Bulletin 20*(4), pp. 33–37, 1989.

[Molich & Nielsen, 1990] R. Molich and J. Nielsen. Improving a Human-Computer Dialogue. *Communications of the ACM 33*(3), pp. 338–348, 1990.

[Moore, 1985] J. D. Moore. *A Reactive Approach to Explanation in Expert and Advice Giving Systems.* Dissertation at the University of California, Los Angeles, 1985.

[Moore & Swartout, 1989] J. Moore and W. Swartout. A Reactive Approach to Explanation. In: *Proc. IJCAI-89*, pp. 1504–1510 (Detroit, MI, Aug. 20–25, 1989). Morgan Kaufmann, San Mateo, CA, 1989.

[Morik, 1989] K. Morik. Sloppy Modeling. In: K. Morik, editor, *Knowledge Representation and Organization in Machine Learning*, pp. 107–134. Springer-Verlag, New York, 1989.

[Morik, 1991] K. Morik. Underlying Assumptions of Knowledge Acquisition and Machine Learning. *Knowledge Acquisition Journal 3*(2), pp. 137–156, 1991.

[Morik, 1993] K. Morik. Balanced Cooperative Modeling. *Machine Learning 10*(1), pp. 217–235, 1993. Revised version of the paper: R. S. Michalski and G. Tecuci, editors, *Proceedings of the First International Workshop on Multistrategy Learning (MSL-91)*, pp. 65–80. George Mason University, Fairfax, VA, 1991.

[Morik et al., 1993] K. Morik, S. Wrobel, J.-U. Kietz, and W. Emde. *Knowledge Acquisition and Machine Learning: Theory Methods and Applications.* Academic Press, New York, 1993.

[Moser et al., 1991] K. A. Moser, D. J. Mazzola, R. T. Keim, and A. S. Philippakis. Modeling the Information Systems Architecture: An Object-Oriented Approach. In: *Proceedings of the Twenty-Fourth Annual Hawaii International Conference on*

System Science, vol. IV, pp. 83–92 (Kauai, HI, Jan. 8–11, 1991). IEEE Computer Society Press, Washington, DC, 1991.

[Muggleton & Feng, 1990] S. Muggleton and C. Feng. Efficient Induction of Logic Programs. In: *Proceedings of the First Conference on Algorithmic Learning Theory.* Ohmsha Publishers, Tokyo, 1990.

[Müller, 1991] B. S. Müller. *Degrees of Cannedness.* DIAMOD Report 11, GMD, Sankt Augustin, November 1991. Also in: B. Becker and T. Gordon, editors, *AI Reader '93.* Arbeitspapiere der GMD 749, GMD, Sankt Augustin, 1993.

[Müller, 1992] B. S. Müller. *Literarische Rhetorik und Text-Generierung—Eine Skizze.* DIAMOD Report 19, GMD, Sankt Augustin, Dec. 1992.

[Müller & Becker, 1991] B. S. Müller and A. Becker. *Sprechende Namen, unbenannte Relationen, unerklärbare Bodies—Schwierigkeiten beim Erklären von Programmen.* DIAMOD Report 12, GMD, Sankt Augustin, Dec. 1991.

[Müller & Sprenger, 1991] B. S. Müller and M. Sprenger. *Dialogabhängige Erklärungsstrategien für modellbasierte Expertensysteme—Das Projekt DIAMOD.* DIAMOD Report 2, GMD, Sankt Augustin, July 1991.

[Myers, 1990] B. A. Myers. A New Model for Handling Input. *ACM Transactions on Information Systems 8*(3), pp. 289–320, 1990.

[Myers & Rosson, 1992] B. Myers and M. Rosson. Survey on User Interface Programming. In: P. Bauersfeld, J. Bennett, and G. Lynch, editors, *Proc. CHI'92,* pp. 195–202 (Monterey, CA, May 3–7, 1992). Addison-Wesley, Reading, MA, 1992.

[Myers et al., 1990] B. A. Myers, D. Giuse, R. B. Dannenberg, B. van der Zanden, D. Kosbie, E. Pervin, A. Mickish, and P. Marchal. Garnet: Comprehensive Support for Graphical, Highly Interactive User Interfaces. *IEEE Computer,* pp. 71–85, Nov. 1990.

[Nagel & Newman, 1959] E. Nagel and J. R. Newman. *Gödel's Proof.* Routledge & Kegan Paul Ltd., London, 1959.

[Narayanan & Chandrasekaran, 1991] N. H. Narayanan and B. Chandrasekaran. Reasoning Visually about Spatial Interactions. In: *Proc. IJCAI-91,* vol. 1, pp. 360–365 (Darling Harbour, Sydney, Australia, Aug. 24–30, 1991). Morgan Kaufmann, San Mateo, CA, 1991.

[Newell, 1982] A. Newell. The Knowledge Level. *Artificial Intelligence 18,* pp. 82–127, 1982.

[Nielsen, 1990] J. Nielsen. *Hypertext/Hypermedia.* Academic Press, Boston, MA, 1990.

[Nilsson, 1986] N. J. Nilsson. Probabilistic Logic. *Journal of Artificial Intelligence 28*(1), pp. 71–87, 1986.

[Nonogaki & Ueda, 1991] H. Nonogaki and H. Ueda. FRIEND21 Project. A Construction of 21st Century Human Interface. In: S. P. Robertson, G. M. Olson, and J. S. Olson, editors, *Reaching through Technology. Human Factors in Computing Systems. Proc. CHI'91,* pp. 407–417 (New Orleans, LA, April 28–May 2, 1991). Addison-Wesley, Reading, MA, 1991.

[Norcio & Stanley, 1989] A. F. Norcio and J. Stanley. Adaptive Human–Computer Interfaces: A Literature Survey and Perspective. *IEEE Transactions on Systems, Man, and Cybernetics 19*(2), pp. 399–408, 1989.

[O'Malley et al., 1985] C. E. O. O'Malley, S. W. Draper, and M. S. Riley. Constructive Interaction: A Method for Studying Human–Computer–Human Interaction. In: B. Shackel, editor, *Proc. INTERACT'84,* pp. 269–274 (London, 1984). Elsevier Science Publishers, Amsterdam, 1985.

[Oppermann, 1994a] R. Oppermann. Adaptively Supported Adaptability. *International Journal of Human-Computer Studies 40,* pp. 455–472, 1994.

[Oppermann, 1994b] R. Oppermann, editor. *Adaptive User Support*. Lawrence Erlbaum Associates, Hillsdale, NJ, 1994.

[Orcutt et al., 1986] G. Orcutt, J. Merz, and H. Quinke, editors. *Microanalytic Simulation Models to Support Social and Financial Policy*. North Holland, Amsterdam, 1986.

[OSF, 1993] Open Software Foundation. *OSF/MOTIF Style Guide, Revision 1.2*. Prentice-Hall, London, 1993.

[Paaß, 1992] G. Paaß. *Representation of Multiple Objects for Associative Geometric Reasoning*. TASSO Report 35, GMD, Sankt Augustin, Jan. 1992.

[Palmer, 1983] S. E. Palmer. The Psychology of Perceptual Organization: A Transformational Approach. In: J. Beck, editor, *Human and Machine Vision*, pp. 269–339. Academic Press, Orlando, FL, 1983.

[Palmiter & Elkerton, 1991] S. Palmiter and J. Elkerton. An Evaluation of Animated Demonstrations for Learning Computer-Based Tasks. In: S. P. Robertson, G. M. Olson, and J. S. Olson, editors, *Reaching through Technology. Human Factors in Computing Systems. Proc. CHI'91*, pp. 257–263 (New Orleans, LA, April 28–May 2, 1991). Addison-Wesley, Reading, MA, 1991.

[Pankoke-Babatz, 1989] U. Pankoke-Babatz, editor. *Computer-Based Group Communication: The AMIGO Activity Model*. Information Technology Series, Ellis Horwood, Chichester, UK, 1989.

[Paris, 1989] C. L. Paris. The Use of Explicit User Models in a Generation System for Tailoring Answers to the User's Level of Expertise. In: A. Kobsa and W. Wahlster, editors, *User Models in Dialog Systems*, pp. 200–232. Springer-Verlag, Berlin, 1989.

[Pavlidis & van Wyk, 1985] T. Pavlidis and C. J. van Wyk. An Automatic Beautifier for Drawings and Illustrations. *Computer Graphics 19*, pp. 225–234, 1985.

[Perlis, 1985] D. Perlis. Languages with Self-Reference {I}: Foundations. *Artificial Intelligence 25*, pp. 301–322, 1985.

[Peters, 1993] K. Peters. *Generierung natürlichsprachlicher Erklärungen in DIAMOD*. DIAMOD Report 24, Universität Bielefeld, Fakultät für Linguistik und Literaturwissenschaft, Bielefeld, Feb. 1993.

[Phillips et al., 1988] M. D. Phillips, H. S. Bashinski, H. L. Ammerman, and C. M. Fligg. A Task-Analytic Approach to Dialogue Design. In: M. Helander, editor, *Handbook of Human–Computer Interaction*, pp. 835–857. Elsevier, Amsterdam, 1988.

[Piatetsky-Shapiro, 1991] G. Piatetsky-Shapiro. Discovery, Analysis, and Presentation of Strong Rules. In: G. Piatetsky-Shapiro and W. Frawley, editors, *Knowledge Discovery in Databases*, pp. 329–348. MIT Press, Cambridge, MA, 1991.

[Pollock, 1988] J. Pollock. Defeasible Reasoning. *Cognitive Science 11*, pp. 481–518, 1988.

[Prinz, 1990] W. Prinz. *Application of the X.500 Directory by Office Systems*. Berichte der GMD 181, R. Oldenbourg Verlag, München, 1990.

[Prinz & Pennelli, 1992] W. Prinz and P. Pennelli. Relevance of the X.500 Directory to CSCW Applications. In: D. Marca and G. Bock, editors, *Groupware: Software for Computer Supported Cooperative Work*, pp. 209–225. IEEE Computer Society Press, Los Alamitos, CA, 1992.

[Quast, 1993] K. J. Quast. Plan Recognition for Context-Sensitive help. In: *Proceedings of the 1993 International Workshop on Intelligent User Interfaces*, pp. 89–96 (Orlando, FL, Jan. 4–7, 1993). ACM Press, New York, 1993.

[Quinlan, 1990] J. R. Quinlan. Learning Logical Definitions from Relations. *Machine Learning 5*(3), pp. 239–266, 1990.

[Rehbein, 1980] J. Rehbein. *Hervorlocken, Verbessern, Aneignen: Diskursanalytische Studien des Fremdsprachenunterrichts.* Mimeo, Bochum, 1980.

[Reimer, 1989] U. Reimer. *FRM: Ein Frame-Repräsentationsmodell und seine formale Semantik.* Springer-Verlag, Berlin, 1989.

[Reiterer, 1993] H. Reiterer. The Development of Design Aid Tools for a Human Factor Based User Interface Design. In: *Proceedings of the 1993 IEEE International Conference on Systems, Man and Cybernetics,* vol. 4, pp. 361–366. IEEE, Piscataway, NJ, 1993.

[Reiterer, 1994] H. Reiterer. An Enabling System for User Interface Design. In: S. Robertson, editor, *Contemporary ERGONOMICS 1994, Proceedings of the Ergonomic Society's 1994 Annual Conference,* pp. 125–130. Taylor & Francis, London, 1994.

[Retz-Schmidt, 1988] G. Retz-Schmidt. Various Views on Spatial Prepositions. *AI Magazine,* Summer 1988, pp. 95–105, 1988.

[Robinson, 1993] M. Robinson. Keyracks and Computers: An Introduction to "Common Artefact" in Computer Supported Cooperative Work (CSCW). *Wirtschaftsinformatik* 35(2), pp. 157–166, 1993.

[Robinson & Bannon, 1991] M. Robinson and L. Bannon. Questioning Representations. In: L. Bannon, M. Robinson, and K. Schmidt, editors, *ECSCW'91, Proceedings of the Second European Conference on Computer-Supported Cooperative Work,* pp. 219–234 (Amsterdam, Sept. 24–27, 1991). Kluwer, Amsterdam, 1991.

[Rock & Palmer, 1990] I. Rock and S. E. Palmer. The Legacy of Gestalt Psychology. *Scientific American 12/1990,* pp. 48–61, 1990.

[Rogers, 1990] R. A. Rogers. *Visions Dancing in Engineer's Heads. AT & T's Quest to Fulfill the Leitbild of a Universial Telephone Service.* WZB-Paper FS II 90–102, Wissenschaftszentrum Berlin, Berlin, 1990.

[Rome, 1991] E. Rome. *EPICT—Benutzerhandbuch V0.1.* TASSO Report 15, GMD, Sankt Augustin, 1991.

[Rome, 1993] E. Rome. MAX, ein maschinelles Gestalt-Erkennungssystem. *KI—Künstliche Intelligenz* 7 (Sonderheft), pp. 70–71, 1993.

[Rome, 1994] E. Rome. Von der Gestalt-Gruppierung zur Gestalt-Erkennung. In: *Proceedings of the First Workshop on Visual Computing.* FhG-IDG, Darmstadt, March 1994.

[Rome et al., 1990] E. Rome, K. Wittur, and D. Bolz. *EPICT—Eine erweiterbare Grafik-Beschreibungssprache.* TASSO Report 4, GMD, Sankt Augustin, 1990.

[Root, 1988] R. W. Root. Design of a Multi-Media Vehicle for Social Browsing. In: *Proceedings of the Conference on Computer-Supported Cooperative Work,* pp. 25–38 (Portland, OR, Sept. 1988). ACM Press, New York, 1988.

[Rose et al., 1991] M. T. Rose, J. P. Onions, and C. J. Robbins. *The ISO Development Environment: User's Manual. Version 6.24,* vols. 1–5, 1991.

[Rosenbloom et al., 1988] P. Rosenbloom, J. Laird, and A. Newell. Meta-Levels in SOAR. In: *Meta-Level Architectures and Reflection,* pp. 227–240. North Holland, Amsterdam, 1988.

[Rosenfeld, 1986] A. Rosenfeld. Pyramid Algorithms for Perceptual Organization. In: *Behavior Research Methods, Instruments, & Computers 18(6),* pp. 595–600, 1986.

[Rowe et al., 1991] L. A. Rowe, J. A. Konstan, B. C. Smith, S. Seitz, and C. Liu. The PICASSO Application Framework. In: *Proceedings of ACM Symposium on User Interface Software and Technology,* pp. 95–105 (Hilton Head, SC, Nov. 11–13, 1991). ACM Press, New York, 1991.

[Russell & Zilberstein, 1991] S. J. Russell and S. Zilberstein. Composing Real-Time Systems. In: *Proc. IJCAI-91,* vol. 1, pp. 212–217 (Darling Harbour, Sydney, Australia, Aug. 24–30, 1991). Morgan Kaufmann, San Mateo, CA, 1991.

[Rutten & Hertzberg, 1993] E. Rutten and J. Hertzberg. Temporal Planner = Nonlinear Planner + Time Map Manager. *AI Communications 6*(1), pp. 18–26, 1993.

[Sacerdoti, 1977] E. D. Sacerdoti. *A Structure for Plans and Behavior.* Elsevier/ North Holland, Amsterdam, 1977.

[Salvendy, 1991] G. Salvendy. Design of Adaptive Interfaces and Flexible Mass Production of Knowledge–Based Systems. In: H.-J. Bullinger, editor, *Human Aspects in Computing. Design and Use of Interactive Systems and Work with Terminals,* pp. 55–68. Elsevier, Amsterdam, 1991.

[Sandewall, 1992] E. Sandewall. *Features and Fluents. A Systematic Approach to the Representation of Knowledge about Dynamical Systems* (Technical Report LiTH-IDA-R-92-30). Dept. of Computer and Information Sciences, Linköping University, Linköping, Sweden, 1992.

[Scheifler & Gettys, 1986] R. W. Scheifler and J. Gettys. The X Window System, *ACM Transactions on Graphics 5*(2), pp. 79–109, 1986.

[Schmidt, 1991] K. Schmidt. Riding a Tiger, or Computer Supported Cooperative Work. In: L. Bannon, M. Robinson, and K. Schmidt, editors, *ECSCW'91, Proceedings of the 2nd European Conference on Computer-Supported Cooperative Work,* pp. 1–16. Kluwer, Dordrecht, 1991.

[Schmucker, 1986] K. J. Schmucker. *Object-Oriented Programming for the Macintosh.* Hayden Book Company, Hasbrouck Heights, NJ, 1986.

[Schreiber et al., 1991] G. Schreiber, B. Bartsch-Spörl, B. Bredeweg, F. van Harmelen, W. Karbach, M. Reinders, E. Vinkhuyzen, and A. Voß. *Designing Architectures for Knowledge-Level Reflection.* ESPRIT Basic Research Action P3178 REFLECT, Deliverable IR.4 RFL/UvA/III.1/4, REFLECT Consortium, University of Amsterdam, Amsterdam, Aug. 1991.

[Schreiber et al., 1993] G. Schreiber, B. Wielinga, and J. Breuker, editors. *KADS: A Principled Approach to Knowledge-Based System Development.* Academic Press, London, 1993.

[Schuler & Smith, 1990] W. Schuler and J. B. Smith. Author's Argumentation Assistant (AAA): A Hypertext-Based Authoring Tool for Argumentative Texts. In: A. Rizk, N. Streitz, and J. Andre, editors, *Hypertext: Concepts, Systems and Applications.* Cambridge University Press, Cambridge, 1990.

[Shapiro, 1983] E. Y. Shapiro. *Algorithmic Program Debugging.* ACM Distinguished Doctoral Dissertations. MIT Press, Cambridge, MA, 1983.

[Simari & Loui, 1992] G. R. Simari and R. P. Loui. A Mathematical Treatment of Defeasible Reasoning and Its Implementation. *Artificial Intelligence 53*(2–3), pp. 125–157, 1992.

[Singh et al., 1990] G. Singh, C. H. Kok, and T. Y. Ngan. Druid: A System for Demonstrational Rapid User Interface Development. *Proceedings of the ACM SIGGRAPH Symposium on User Interface Software and Technology,* pp. 167–177 (Snowbird, UT, Oct. 3–5, 1990). ACM Press, New York, 1990.

[Smith, 1982] B. Smith. Reflection and Semantics in a Procedural Language (Technical Report TR-272). MIT, Computer Science Lab., Cambridge, MA, 1982. Also in: R. J. Brachman and H. J. Levesque, editors, *Readings in Knowledge Representation,* pp. 31–40. Morgan Kaufman, CA, 1985.

[Snyder & Lynch, 1991] J. M. Snyder and K. J. Lynch. An Overview of CARAT: A Computer Assisted Research and Analysis Tool. In: J. F. Nunamaker and R. H. Sprague, editors, *Proceedings of the Twenty-Fourth Annual Hawaii International Conference on System Science,* vol. IV, pp. 343–353 (Kauai, HI, Jan. 8–11, 1991). IEEE Computer Society Press, Washington, DC, 1991.

[Sohlenkamp & Chwelos, 1994] M. Sohlenkamp and G. Chwelos. Integrating Communication, Cooperation, and Awareness: The DIVA Virtual Office Environment. In: *Proceedings of the ACM 1994 Conference on Computer Supported Cooperative Work (CSCW'94),* pp. 331–343, (Chapel Hill, NC, Oct. 22–26). ACM Press, New York, 1994.

[Sommer, 1993] E. Sommer. Cooperation of Data-Driven and Model-Based Induction Methods for Relational Learning. In: R. S. Michalski and G. Tecuci, editors, *Second International Workshop on Multistrategy Learning (MSL-93),* pp. 180–187 (Harpers Ferry, WV, 1993). George Mason University, Fairfax, VA, 1993.

[Sommer, 1994a] E. Sommer. Learning Relations Without Closing the World. In: F. Bergadano and L. De Raedt, editors, *Machine Learning: ECML-94, European Conference on Machine Learning,* pp. 419–422 (Catania, Italy, April 6–8, 1994). Springer-Verlag, Berlin, 1994.

[Sommer, 1994b] E. Sommer. Rulebase Stratification: An Approach to Theory Restructuring. In: S. Wrobel, editor, *Proceedings of the ECML MLNet Familiarization Workshop on Theory Revision and Restructuring in Machine Learning,* pp. 356–360 (Catania, Italy, April 10, 1994). Arbeitspapiere der GMD 824, GMD, Sankt Augustin, 1994.

[Sommer et al., 1994] E. Sommer, K. Morik, J.-M. Andre, and M. Uszynski. *What Online Machine Learning can do for Knowledge Acquisition—A Case Study.* Arbeitspapiere der GMD 757, GMD, Sankt Augustin, 1994.

[Spenke, 1993] M. Spenke. From Undo to Multi-User Applications—The Demo. In: *Proceedings of the CHI Conference on Human Factors in Computing Systems,* pp. 468–469 (Amsterdam, April 24–29, 1993). ACM Press, New York, 1993.

[Spenke & Beilken, 1990] M. Spenke and C. Beilken. An Overview of GINA—The Generic Interactive Application. In: D. A. Duce, M. R. Gomes, F. R. A. Hopgood, and J. R. Lee, editors, *User Interface Management and Design, Proceedings of the Workshop on User Interface Management Systems and Environments,* pp. 273–293 (Lisbon, Portugal, June 4–6, 1990). Springer-Verlag, Berlin, 1990.

[Spenke et al., 1992a] M. Spenke, C. Beilken, T. Berlage, A. Bäcker, and A. Genau. *GINA User Manual, Version 2.1 for Common Lisp.* Arbeitspapiere der GMD 614, GMD, Sankt Augustin, 1992.

[Spenke et al., 1992b] M. Spenke, C. Beilken, T. Berlage, A. Bäcker, and A. Genau. *GINA Reference Manual, Version 2.1.* Arbeitspapiere der GMD 615, GMD, Sankt Augustin, 1992.

[Spirtes et al., 1993] P. Spirtes, C. Glymour, and R. Scheines. *Causality, Prediction, and Search.* Springer-Verlag, New York, 1993.

[Sprenger, 1992] M. Sprenger. *Explanations for KADS—KADS for Explanations? Evaluating the Explanation Capabilities of a Model Based Approach to Expert System Construction.* Presented at Samos Workshop on Task-Based Explanation (Samos, Greece, 1992). DIAMOD Report 20, GMD, Sankt Augustin, Dec. 1992.

[Sprenger, 1993] M. Sprenger. Explanation Strategies for KADS-Based Expert Systems. In: H. Horacek and M. Zock, editors, *New Concepts in Natural Language Generation: Planning, Realization, and Systems,* pp. 27–56. Pinter, London, 1993.

[Sprenger & Wickler, 1993] M. Sprenger and G. Wickler. DIAMOD: An Explanation Component for the Assignment Probem. In: M. Wick, editor, *Workshop Notes on the*

IJCAI-93 Workshop on Explanation and Problem Solving, (Chambery, France, Aug. 29, 1993). University of Wisconsin, Eau Claire, WI, 1993.

[Steedman, 1990] D. Steedman. *Abstract Syntax Notation One, The Tutorial Reference.* Technology Appraisals, Isleworth, 1990.

[Steels, 1990] L. Steels. Components of Expertise. *AI Magazine 11*(2), pp. 28–49, Summer 1990. Also as: AI Memo 88–16, AI Lab, Free University of Brussels, Brussels.

[Stefik, 1981] M. Stefik. Planning and Meta-Planning (molgen: Part 2). *AI Journal 16,* pp. 141–170, 1981.

[Suchman, 1994] L. Suchman. Do Categories Have Politics? The Language/Action Perspective Reconsidered. *Computer-Supported Cooperative Work 2,* pp. 177–190, 1994.

[Suchman & Trigg, 1991] L. A. Suchman and R. H. Trigg. Understanding Practice: Video as a Medium for Reflection and Design. In: J. Greenbaum and M. Kyng, editors, *Design at Work: Cooperative Design of Computer Systems,* pp. 65–90. Lawrence Erlbaum Associates, Hillsdale, NJ, 1991.

[Sussman, 1975] G. J. Sussman. *A Computer Model of Skill Acquisition.* Artificial Intelligence Series, vol. 1, American Elsevier, New York, 1975.

[Swartout, 1992] W. Swartout. Explainable Expert Systems Architectures: Putting the Pieces Together. In: *Proceedings Samos Workshop on Task based Explanation,* (Samos, Greece, 1992). Reprinted in Workshop Explanation (1992), pp. 51–57.

[Tate, 1994] A. Tate. The Emergence of "Standard" Planning and Scheduling System Components—Open Planning and Scheduling Architectures. In: C. Bäckström and E. Sandewall, editors, *Current Trends in AI Planning. EWSP'93—2nd European Workshop on Planning,* pp. 14–32 (Vadstena, Sweden, Dec. 9–11, 1993). IOS Press, Amsterdam, 1994.

[Tenenberg, 1991] J. D. Tenenberg. Abstraction in Planning. In: J. Allen, H. Kautz, R. Pelavin, and J. Tenenberg, editors, *Reasoning about Plans,* chapter 4, pp. 213–283. Morgan Kaufmann, San Mateo, CA, 1991.

[Tepper, 1993] A. Tepper. Future Assessment by Metaphors. In: *Behaviour & Information Technology 12*(6), pp. 336–345, 1993.

[Thaise, 1988] A. Thaise, editor. *From Standard Logic to Logic Programming.* John Wiley & Sons, Chichester, 1988.

[Thiébaux & Hertzberg, 1992] S. Thiébaux and J. Hertzberg. A Semi-Reactive Planner Based on a Possible Models Action Formalization. In: J. Hendler, editor, *Artificial Intelligence Planning Systems: Proceedings of the First International Conference (AIPS-92),* pp. 228–235 (College Park, MD, June 15–17, 1992). Morgan Kaufmann, San Mateo, CA, 1992.

[Thieme, 1989] S. Thieme. The Acquisition of Model Knowledge for a Model-Driven Machine Learning Approach. In: K. Morik, editor, *Knowledge Representation and Organization in Machine Learning,* pp. 177–191. Springer-Verlag, New York, 1989.

[Thomas, 1993] C. G. Thomas. Design, Implementation, and Evaluation of an Adaptive User Interface. *Knowledge-Based Systems 6*(4), Special Issue on Intelligent Interfaces, pp. 230–238, 1993.

[Toulmin, 1958] S. E. Toulmin. *The Uses of Argument.* Cambridge University Press, Cambridge, 1958.

[Touretzky et al., 1988] D. S. Touretzky, J. F. Horty, and R. H. Thomason. A Clash of Intuitions: The Current State of Nonmonotonic Multiple Inheritance Systems. In: *Proc. IJCAI-87,* pp. 476–482 (Milan, Italy, Aug. 23–28, 1987). Morgan Kaufman, Los Altos, CA, 1987.

[Treisman, 1985] A. Treisman. Preattentive Processing in Vision. *Computer Vision, Graphics, and Image Processing 31*, pp. 156–177, 1985.

[Tufte, 1983] E. R. Tufte. *The Visual Display of Quantitative Information*. Graphics Press, Cheshire, CT, 1983.

[UIMS Tool Developers Workshop, 1992] UIMS Tool Developers Workshop, A Meta-model for the Runtime Architecture of an Interactive System. *SIGCHI Bulletin 24*(1), pp. 32–37, 1992.

[Ulich, 1978] E. Ulich. Über das Prinzip der differentiellen Arbeitsgestaltung. *Industrielle Organisation 47*, pp. 566–568, 1978.

[Umesh, 1988] R. M. Umesh. A Technique for Cluster Formation. *Pattern Recognition 21*(4), pp. 393–400, 1988.

[Uthurusamy et al., 1991] R. Uthurusamy, U. Fayyad, and S. Spangler. Learning Useful Rules from Inconclusive Data. In: G. Piatetsky-Shapiro and W. Frawley, editors, *Knowledge Discovery in Databases*, pp. 141–158. MIT Press, Cambridge, MA, 1991.

[van der Veer et al., 1985] G. van der Veer, M. Tauber, Y. Waern, and B. van Muylwijk. On the Interaction between System and User Characteristics. *Behaviour and Information Technology 4*(4), pp. 289–308, 1985.

[van Harmelen, 1991] F. van Harmelen. *Meta-Level Inference Systems*. Research Notes in AI. Morgan Kaufmann, San Mateo, CA, 1991.

[Vellum, 1992] see [Ashlar .i.Vellum;, 1992]

[Vlissides & Linton, 1990] J. M. Vlissides and M. A. Linton. Unidraw: A Framework for Building Domain-Specific Graphical Editors. *ACM Transactions on Information Systems 8*(3), pp. 237–268, 1990.

[Voß, 1994] A. Voß, editor. *Similarity Concepts and Retrieval Methods*. FABEL Report 13, GMD, Sankt Augustin, Feb. 1994.

[Voß et al., 1990] A. Voß, W. Karbach, U. Drouven, and D. Lorek. Competence Assessment in Configuration Tasks. In: L. C. Aiello, editor, *Proceedings of the 9th European Conference on Artificial Intelligence (ECAI-90)*, pp. 676–681 (Stockholm, Sweden, Aug. 6–10, 1990). ECCAI, Pitman, London, 1990.

[Voß et al., 1991] A. Voß, W. Karbach, B. Bartsch-Spörl, and B. Bredeweg. Reflection and Competent Problem Solving. In: T. Christaller, editor, *GWAI-91, 15th German Workshop on Artificial Intelligence*, pp. 206–215 (Bonn, Dec. 16–20, 1991). Springer-Verlag, Berlin, 1991.

[Voß et al., 1992] A. Voß, W. Karbach, C. H. Coulon, U. Drouven, and B. Bartsch-Spörl. Generic Specialists in Competent Behavior. In: B. Neumann, editor, *Proc. ECAI-92*, pp. 567–571 (Vienna, Austria, Aug. 3–7, 1992). Wiley, Chichester, 1992.

[Wahlster, 1991] W. Wahlster. User and Discourse Models for Multimodal Communication. In: J. W. Sullivan and S. W. Tyler, editors, *Intelligent User Interfaces*, pp. 45–67. Addison-Wesley, Reading, MA, 1991.

[Walther et al., 1992] J. Walther, A. Voß, M. Linster, T. Hemmann, H. Voß, and W. Karbach. *Momo*. Arbeitspapiere der GMD 658, GMD, Sankt Augustin, 1992.

[Way, 1991] E. C. Way. *Knowledge Representation and Metaphor*. Kluwer, Dordrecht, 1991.

[Webster, 1989] B. F. Webster. *The NeXT Book*. Addison-Wesley, Reading, MA, 1989.

[Weinand et al., 1989] A. Weinand, E. Gamma, and R. Marty. Design and Implementation of ET++, a Seamless Object-Oriented Application Framework. *Structured Programming 10*(2), pp. 63–87, 1989.

[Wertheimer, 1923] M. Wertheimer. Untersuchungen zur Lehre von der Gestalt II. *Psychologische Forschung 4*, pp. 301–350, 1923.

[Wick & Thompson, 1992]   M. R. Wick and W. B. Thompson. Reconstructive Expert System Explanation. *Artificial Intelligence 54*(1–2), pp. 33–70, 1992.

[Wickler & Heider, 1992]   G. Wickler and S. Heider. Reducing Traces for Answering Dynamic Questions. Paper presented at *Samos Workshop on Task-Based Explanation* (Samos, Greece, 1992). DIAMOD Report 14, GMD, Sankt Augustin, August 1992.

[Wielinga & Breuker, 1986]   B. Wielinga and J. Breuker. Models of Expertise. In: *Proc. ECAI-86*, pp. 306–318 (Brighton, UK, July 21–25, 1986). ECCAI, 1986.

[Wielinga & Swartout, 1992]   B. Wielinga and W. Swartout. The Table Comparing KADS and EES as introduced by B. Wielinga with corrections by W. Swartout. In: Workshop Explanation (1992), pp. 59–62.

[Wielinga et al., 1991]   B. Wielinga, A. Schreiber, and J. Breuker. *KADS: A Modelling Approach to Knowledge Engineering.* Research Report ESPRIT-Project P5248 KADS-II/T1.1/PP/UvA/008/1.0, University of Amsterdam, Amsterdam, 1991.

[Wilensky, 1980]   R. Wilensky. Meta-Planning. In: R. Wilensky, editor, *Proc. AAAI-80*, pp. 334–336 (Stanford, CA, 1980). AAAI, Menlo Park, CA, 1980.

[Winograd, 1994]   T. Winograd. Categories, Disciplines, and Social Coordination. *Computer-Supported Cooperative Work 2*, pp. 191–197, 1994.

[Winograd & Flores, 1986]   T. Winograd and F. Flores. *Understanding Computers and Cognition, A New Foundation for Design.* Ablex, Norwood, NJ, 1986.

[Workshop Explanation, 1992]   *Explanation Facilities for Model-Based Expert Systems.* Materials from the Workshop held on May 25 and 26, 1992 in Sankt Augustin, Germany. Internal Report, Project DIAMOD. GMD, Sankt Augustin, Dec. 1992.

[Wrobel, 1988]   S. Wrobel. Automatic Representation Adjustment in an Observational Discovery System. In: D. Sleeman, editor, *Proceedings of the 3rd European Working Session on Learning*, pp. 253–262 (Glasgow, Scotland, Oct. 3–5, 1988). Pitman, London, 1988.

[Wrobel, 1993]   S. Wrobel. On the Proper Definition of Minimality in Specialization and Theory Revision. In: *Proceedings of the Sixth European Conference on Machine Learning (ECML-93)*, pp. 65–82 (Vienna, Austria, April 5–7, 1993). Springer-Verlag, Berlin, 1993. Also as: Arbeitspapiere der GMD 730, GMD, Sankt Augustin, 1993.

[Wrobel, 1994a]   S. Wrobel. Concept Formation During Interactive Theory Revision. *Machine Learning 14*, pp. 169–191, 1994.

[Wrobel, 1994b]   S. Wrobel. *Concept Formation and Knowledge Revision.* Kluwer, Dordrecht, 1994.

[X.500, 1993]   X.500, *The Directory. CCITT Rec. X.500-X.521 | ISO/IEC Standard 9594*, 1993.

[Zachman, 1987]   J. A. Zachman. A Framework for Information Systems Architecture. *IBM Systems Journal 26*(3), pp. 276–292, 1987.

[Zelazny, 1986]   G. Zelazny. *Wie aus Zahlen Bilder werden.* Gabler Verlag, Wiesbaden, 1986.

[Zülch & Starringer, 1984]   G. Zülch and M. Starringer. Differentielle Arbeitsgestaltung in der Fertigung für elektronische Flachbaugruppen. *Zeitschrift für Arbeitswissenschaft 38*, pp. 211–216, 1984.

Subject Index

Author Index

Printed and bound by CPI Group (UK) Ltd, Croydon, CR0 4YY

17/10/2024

01775683-0005